GRAND COMPLICATIONS®

VOL. IX

WATCHES INTERNATIONAL®

THE ORIGINAL ANNUAL OF THE WORLD'S WATCH COMPLICATIONS AND MANUFACTURERS®

tourbillon international®

GRAND COMPLICATIONS®

THE ORIGINAL ANNUAL OF THE WORLD'S WATCH COMPLICATIONS AND MANUFACTURERS®

SPECIAL MINUTE REPEATER EDITION

First published in the United States in 2013 by

TOURBILLON INTERNATIONAL
A MODERN LUXURY COMPANY

11 West 25th Street, 8th Floor
New York, NY 10010
Tel: +1 (212) 627-7732 Fax +1 (312) 274-8418
www.modernluxury.com/watches

In association with RIZZOLI INTERNATIONAL PUBLICATIONS, INC.

300 Park Avenue South, New York, NY 10010

ISBN: 978-0-8478-4034-2

PRINTED IN ITALY

COVER: KANTHAROS (CHRISTOPHE CLARET)

RICHARD MILLE
A RACING MACHINE ON THE WRIST
RM 007
trusted.com™
TRUST WHAT YOU BUY
© Renaud Cortouer
www.richardmille.com

TOURBILLON SPLIT SECONDS COMPETITION CHRONOGRAPH RM 050 FELIPE MASSA

Manual winding tourbillon movement, lightened to extreme
Power reserve: circa 70 hours
Split seconds chronograph
(column wheels in titanium)
Split seconds mechanism with improved performance
Spline screws in grade 5 titanium for the bridges
Torque indicator
Function indicator
Number of jewels: 35
Balance wheel: GLUCYDUR, with 2 arms and 4 setting screws
Moment of inertia 10 mg·cm^2, angle of lift 53°
Frequency: 21,600 vph (3 Hz)
Case in carbon nanotubes
Torque limiting crown
Extreme finishing for the movement
Hand-polished bevelling
Hand-polished locking-sections
Microblasted milled sections
Lapped and polished contact points
Burnished pivots

Letter from the President

Timely Harmonies

It is my great pleasure to introduce the 2013 edition of *Grand Complications*, with a special focus on minute repeater watches. In many ways, the minute repeater is the quintessential luxury watch complication: not only do audible time indications predate even the simplest visual ones, but the chimes that signify the hour, quarter-hour and minute may seem superfluous in this age of endless beeping devices. However, the exceptional artistry and skill that go into every minute repeater model—the appeal to the ear, as well as the eye and the skin—are reflections of the imagination and meticulous attention to detail throughout haute horology.

With recognition and respect for the purely mechanical realm, Tourbillon International also continues a bold foray into the digital world, following the overwhelming success of our apps for iPad, iPhone and Android. We invite you to download the *Grand Complications* app—just like your favorite timepiece, you will be able to refer to it whenever and wherever you desire.

The in-depth explorations of the world's most challenging complications, the interviews with leading industry figures and the full-page photographs of exquisite timepieces from the world's most respected horologers make this hardcover edition of *Grand Complications* a resource that you will want to return to in admiration—again, just like your favorite timepiece.

Michael Dickey

PROJECT Z6 BLACK EDITION
ALARM MOVEMENT - 72 HOURS POWER RESERVE
ZALIUM™ EXCLUSIVE ALLOY

HARRY WINSTON®

REINVENTING TIME™

HARRYWINSTON.COM

cellini

FOR 35 YEARS,

Cellini Jewelers has earned the respect of aficionados who share its passion for mechanical timepieces by offering one of the most extensive collections of the world's best watches.

That incredible depth is what separates Cellini, and allows it to deliver an unparalleled experience to collectors. Between its locations in the Hotel Waldorf-Astoria and on Madison Avenue, Cellini has the ability to satisfy nearly every request, says company president Leon Adams. "It's rare that someone asks for a watch we don't have," he explains. "People value the immediacy of being able to find what they want, when they want it."

What the company does best, perhaps, is use that range to help someone find the right watch. "If you like a particular style—be it a chronograph or an ultra-thin dress watch—we can line up 20 models from different brands so you can judge for yourself how a watch looks and feels," Adams says.

It's impossible to get that kind of brand-versus-brand comparison and neutral opinion at a store dedicated to a single watchmaker. "Our views have been shaped from handling so many brands for so long," he explains. "People trust our advice because we strive to be accurate and objective."

▲ **DROP EARRINGS**

Drop earrings with more than 11 carats of rose-cut fancy yellow diamonds.

◄ **HYT H1**

H1 in titanium coated entirely in black DLC by HYT.

▼ **PEARL BANGLES**

Stackable South Sea pearl bangles featuring emeralds, rubies and sapphires, accented with diamonds in 18-karat gold.

DEEP ROOTS

In the late '70s and early '80s, the popularity of mass-market quartz watches came at the expense of companies producing mechanical wristwatches. As a result, Cellini emerged as one of the few places in America where it was still possible to find the handiwork of such historic firms as Audemars Piguet, IWC, Jaeger-LeCoultre and Vacheron Constantin.

As the mechanical watch's popularity began to soar in the early '90s, Cellini became known as the place to see both established and rising brands. The impressive list of watchmakers that found an early home at Cellini includes A. Lange & Söhne, De Bethune, DeWitt, Franck Muller, Hublot, Jean Dunand, Ludovic Ballouard, Maîtres du Temps, Parmigiani, Richard Mille and Roger Dubuis.

Even today, Cellini's role as an influential tastemaker continues to grow as it welcomes two new brands. It now offers a full range of the Bulgari Serpenti, a serpent-shaped bracelet watch for women. Cellini is also home to HYT, a Swiss firm acclaimed for its award-winning debut, the groundbreaking H1.

COLOR ME IMPRESSED

More than just watches, Cellini is also ranked among the finest jewelers in New York City. It added bold splashes of color to its collection this year, unveiling several vibrant new pieces.

Among the standouts is a dramatic emerald and diamond necklace. Together, the emeralds weigh an impressive 59.43 carats. But the color is what really sets this piece apart, says Claudette Levy, Cellini's jewelry buyer. "The greens are gorgeous. And, it's not often that you see emeralds so well-matched in a necklace this size," she explains.

Another bright spot in this colorful display is a pair of earrings with rose-cut fancy yellow diamonds edged with white diamonds. Levy says the inverted teardrop shape accentuates the face nicely. "The curves really follow the natural contours to emphasize your features," she says.

Cellini also introduced a colorful twist on its South Sea Pearl bangles by adding rubies, sapphires or emeralds. "The rich, natural shade of each gemstone looks crisp against the backdrop of the white diamond rondelles and 18-karat white-gold setting," Levy says.

Unrivaled selection, sublime quality and impeccable service all ensure Cellini's reputation as a luxury landmark for years to come.

ROTONDE DE CARTIER FLYING TOURBILLON
(Cartier)

MOSER PERPETUAL MOON
(H. Moser & Cie)

TONDA RETROGRADE ANNUAL CALENDAR
(Parmigiani)

PULSION FLYING TOURBILLON SKELETON
(Roger Dubuis)

Hotel Waldorf-Astoria
301 Park Avenue at 50th Street
New York, NY 10022
212-751-9824

509 Madison Avenue at 53rd Street
New York, NY 10022
212-888-0505

800-CELLINI
www.CelliniJewelers.com

◂ **SERPENTI**
Serpenti Jewelry with diamond bracelet in white gold by Bulgari.

▴ **DROP NECKLACE**
Emerald and diamond necklace with a double row of 25 oval-shaped emeralds totaling nearly 60 carats.

▸ **ROYAL OAK TOURBILLON**
Extra-Thin Royal Oak Tourbillon in rose gold by Audemars Piguet.

BRANDS CARRIED

A. LANGE & SÖHNE
AUDEMARS PIGUET
BACKES & STRAUSS
BELL & ROSS
BULGARI
CARTIER
CHOPARD
DE BETHUNE
DEWITT
FRANCK MULLER
GIRARD-PERREGAUX
H. MOSER & CIE.
HUBLOT
HYT
IWC
JAEGER-LECOULTRE
JEAN DUNAND
LUDOVIC BALLOUARD
MAÎTRES DU TEMPS
PARMIGIANI
PIAGET
RICHARD MILLE
ROGER DUBUIS
ULYSSE NARDIN
VACHERON CONSTANTIN
ZENITH

CHANEL

Letter from the Publisher

I believe I can fly

In the 1980s, showing off a prancing horse or a leaping jaguar made one not only a marriage-worthy individual, but also the lucky owner of an automobile—or fleet thereof—that acted as status symbol. In 2013, the most impressive mechanics can now be displayed on the wrist. Just like a car, a watch has become an object of recognition. It can organize individuals into modern tribes, and allows everyone to signal their belonging in a particular group. From the comfortable collector to the penniless youngster, from Piaget to IWC, Zenith or Chopard, each band of aficionados goes to bat for its own team. Purists claim their role in the "noble" branch of the field, and largely wear creations by Patek Philippe, Vacheron Constantin, Blancpain or Breguet. Financial sharks affirm their "killer" reputations by wearing daring models from Urwerk, MB&F or Hautlence, symbols of social success as well as financial triumph. Highly attuned to symbolic values, the world of show business casts its vote for the legendary timepieces from the biggest houses: the Reverso from Jaeger-LeCoultre, the Royal Oak from Audemars Piguet or Cartier's famous Tank. A brand, a watch, and something to believe in!

However, schisms do occur among the horological faithful. For instance, Hubot's Big Bang models show up on the wrists of both businessmen and athletes. Chanel's J12, on the other hand, might be worn by a famous swimmer or supermodel. Even though the highest echelons of athletics might not seem to have much to do with the catwalk, the idea remains the same: everyone wants to wear a model with which he or she identifies completely. Doctrine can splinter into several different sects, which still share the same universal marketing motto. Adoration depends on the cult of difference and distinction, and it is all wrapped up in social status, the idea of belonging to an elite and the entry ticket to a world of dreams. It takes more than a beautiful movement, an original design, and curves worthy of Monica Bellucci to arouse the ardor of a true believer. It takes wings…

Caroline Childers

F.P.JOURNE
Invenit et Fecit
Entirely invented and manufactured in-house
fpjourne.com
F.P.JOURNE
Invenit et Fecit
Remontoir d'Egalité avec Seconde Morte
Grand Prix d'Horlogerie de Genève
Golden Hand Grand Prix • 2004
Tourbillon Souverain - Ref. TN
Remontoire and dead beat second
18K solid Gold movement
Platinum or red Gold case
THE BOUTIQUES
PARIS +33 1 42 68 08 00 NEW YORK +1 212 644 5918 MIAMI +1 305 993 4747 LOS ANGELES +1 310 294 8585
GENEVA +41 22 810 33 33 TOKYO +81 3 5468 0931 BEIJING +86 10 8517 2036 HONG KONG +852 2522 1868

HUBLOT
THE ART OF FUSION

Jean-Claude Biver, Chairman of the Board & Ricardo Guadalupe, CEO

Under NEW Management

IN A CHANGE ANNOUNCED IN JANUARY OF 2012, RICARDO GUADALUPE HAS TAKEN ON THE TITLE OF CHIEF EXECUTIVE OFFICER OF HUBLOT, a new step for a man and a brand, which was nonetheless long in the offing. Ricardo Guadalupe and Jean-Claude Biver, the company's new Chairman of the Board, have been friends and colleagues for over 20 years, and Ricardo Guadalupe's new title reflects an evolution for both men.

Born and raised in Neuchâtel, Ricardo Guadalupe grew up in steeped in watchmaking culture, nurturing a passion for the age-old art. Starting his career as Product Manager at Bulgari, he learned how to expand an iconic jewelry brand into the world of horology by getting involved in every aspect of the business, from design to production to distribution to marketing. Over seven years, he deepened his understanding of the technical and traditional facets of haute horology, as well as an intuitive aesthetic feel for design and creation. After meeting Biver in 1994, Guadalupe joined Blancpain. Following its acquisition by the Swatch Group, the brand needed a visionary or two to rebuild it, and Ricardo Guadalupe and Jean-Claude Biver proved more than up to the task.

Since joining Hublot in 2002, Ricardo Guadalupe has made immeasurable contributions to a brand that is now one of the most popular in the world. Hublot has become known for its iconic, immediately recognizable case shapes, its charismatic brand ambassadors, and its use of unconventional materials. With Ricardo Guadalupe at the helm, Hublot continues its effortless dominance, announcing partnerships such as Ferrari and its Scuderia, the next FIFA World Cup of Football in Brazil 2014, and sponsorships such as 2012 US Basketball Champion Miami Heat and prestigious European soccer clubs Bayern Munich and Juventus Turin.

opposite page: Big Bang Ferrari Magic Gold

Letter from the Editor in Chief

In Search of the Absolute

In the world of complications, it seems as though the watchmaking industry answers to fashionable phenomena, and many houses go for the same trends at the same time. If we consider the number of brands on the market, there is nothing unusual about similarities cropping up in terms of product development. But still. Haute horology is not actually so crowded as to make these coincidences inevitable. It is as if the measurement of time required a well-trained battalion, rather than a few isolated snipers.

These last few years, we have seen a few strong trends that have profoundly influenced what is on offer from mechanical horology. Research into new materials, for example, and silicon in particular, has become one of the major preoccupations of the industry, with each watchmaker taking up the challenge to develop its own alloy with extraordinary properties. This quest has rapidly opened up a new field of research, aimed at addressing the prickly problem of lubrication. Even more recently, high-frequency watches have taken the spotlight, opening up a front against the timing discrepancies that occur with any mechanical watch. We must note that horologers have had amazing results in this domain, especially TAG Heuer, which created a movement that beats at 1,000Hz!

If we take the models presented at the beginning of 2013 as a sign, the newest trend to capture everyone's attention is the constant-force mechanism. After Cabestan (Winch Tourbillon Vertical) and A. Lange & Söhne (Richard Lange Tourbillon "Pour le Mérite"), we are also seeing the fruits of this new research at IWC (Ingenieur Constant-Force Tourbillon), Romain Gauthier (Logical One) and Zenith (Academy Christophe Colomb Hurricane), to cite just a few examples. This answer to the difficulties linked to the loss of isochronism, which finds its origins in watches' fusee-chain transmissions of centuries past, is today notably interpreted by a mechanism that allows the escapement to disengage from the gears for a constant amplitude of the balance-spring's oscillation—a true measure of precision.

Mechanical horology presents a field of investigation limited only by the human mind, and our insatiable desire to master our temporal environment. Even if we don't live second to second, that unit remains the minimal standard for watchmakers, who take any affront to this "code of conduct" as a personal insult. So if they often ride into battle on the same horses, it only proves that this search for the absolute is widely shared.

Michel Jeannot

PANERAI.COM
RADIOMIR
PANERAI
pm
am

GUY ELLIA

www.guyellia.com

MALTE TOURBILLON
(Vacheron Constantin)

CLASSIQUE HORA MUNDI
(Breguet)

BIG PILOT PERPETUAL SAINT-EXUPERY
(IWC)

EXCALIBUR 45
(Roger Dubuis)

OYSTER PERPETUAL SKY-DWELLER
(Rolex)

CHRISTOPHE CLARET

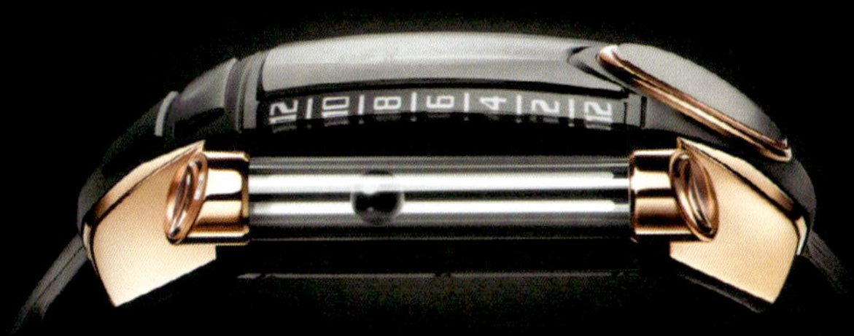

X-TREM-1

PUSHING BACK BOUNDARIES OF HOROLOGY

GRAND COMPLICATIONS®

THE ORIGINAL ANNUAL OF THE WORLD'S WATCH COMPLICATIONS AND MANUFACTURERS®

TOURBILLON INTERNATIONAL
A MODERN LUXURY COMPANY
ADMINISTRATION, ADVERTISING SALES, EDITORIAL, BOOK SALES

11 West 25th Street, 8th Floor • New York, NY 10010
Tel: +1 (212) 627-7732 Fax: +1 (312) 274-8418

Caroline Childers
PUBLISHER

Michel Jeannot
EDITOR IN CHIEF

EDITORS Samson Crouhy
Elise Nussbaum
Julie Singer

ART DIRECTOR Mutsumi Hyuga

CONTRIBUTING EDITORS Fabrice Eschmann
Gregory Pons

TRANSLATIONS Susan Jacquet

COORDINATION Julie Mégevand
Caroline Pita

VICE PRESIDENT OF PRODUCTION Meg Eulberg

VICE PRESIDENT OF MANUFACTURING Sean Bertram

VICE PRESIDENT OF AUDIENCE DEVELOPMENT Eric Holden

DIRECTOR OF INFORMATION TECHNOLOGY Scott Brookman

SALES ADMINISTRATOR Ralph Gago

WEB DISTRIBUTION

www.modernluxury.com/watches

PHOTOGRAPHY

Photographic Archives
Property of Tourbillon International, a Modern Luxury Company

---

MODERN **LUXURY**

Lew Dickey
CHIEF EXECUTIVE OFFICER

Michael Dickey
PRESIDENT

EXECUTIVE VICE PRESIDENT AND CO-CHIEF OPERATING OFFICER John Dickey

EXECUTIVE VICE PRESIDENT AND CO-CHIEF OPERATING OFFICER Jon Pinch

CHIEF FINANCIAL OFFICER JP Hannan

GENERAL COUNSEL Richard Denning

IWC Pilot. Engineered for aviators.

IWC
SCHAFFHAUSEN
SINCE 1868

Perfect fit!

Pilot's Watch Chronograph TOP GUN Miramar. Ref. 3880: An XL uniform doesn't turn a man into a great pilot. After all, the cream of the crop – as members of the TOP GUN squadron are trained to be – regularly go way beyond the limits of ordinary humans. A watch designed to do the same is the Pilot's Watch Chronograph TOP GUN Miramar. Its high-tech ceramic and titanium case, together with a soft-iron inner case to protect the movement against magnetic fields, is absolutely indestructible. Small wonder, then, that it occasionally goes much further than its wearer's wrist. **IWC. Engineered for men.**

Mechanical chronograph movement | Self-winding | 68-hour power reserve when fully wound | Date display | Stopwatch function with minutes and seconds | Flyback function | Small hacking seconds | Soft-iron inner case for protection against magnetic fields (figure) | Screw-in crown | Sapphire glass, convex, antireflective coating on both sides | Water-resistant 6 bar | Ceramic

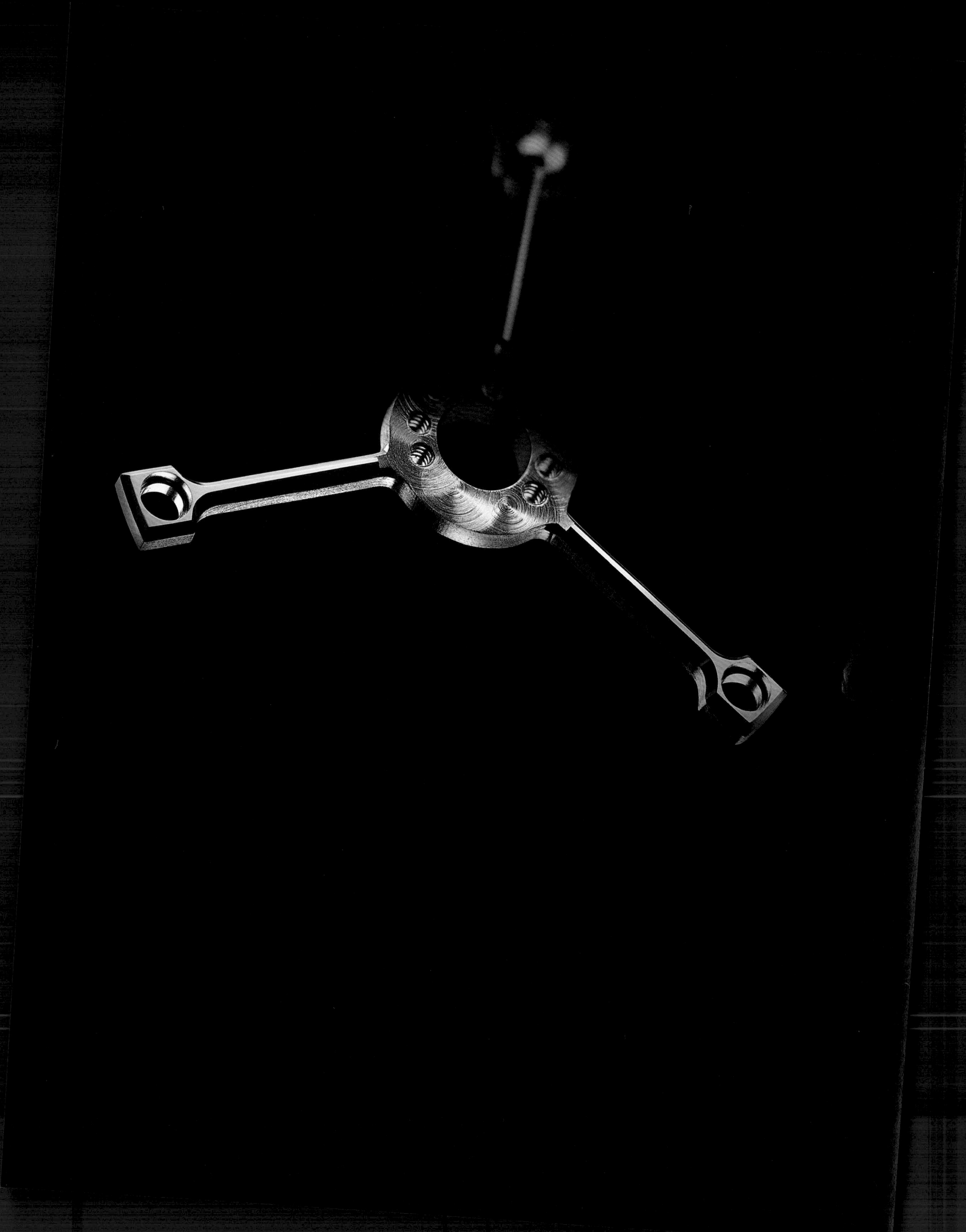

Westime

FOR A QUARTER OF A CENTURY,

Westime has distinguished itself as the ultimate retail destination specializing in extraordinary watches. John Simonian, a third-generation watch expert with a passion for mechanical timepieces, founded Westime in 1987 when he opened the first boutique on Los Angeles's West Side. From its earliest days, Westime catered to a clientele that ranged from Hollywood celebrities and professional athletes, to the region's influential residents and international business travelers. Westime has since earned the return business of discriminating clients from around the globe.

Today, Westime operates three boutiques in Southern California. Westime Beverly Hills is an intimate, multi-level boutique at the heart of the city's most glamorous shopping district. The shop is located on Via Rodeo, a European-style shopping promenade that has emerged as a global destination for the finest watch, jewelry and accessories brands. High ceilings, marble flooring, contemporary furnishings and showcases that feature champagne water relief woodwork complement a precisely edited selection of limited editions, high complications and even custom timepieces created exclusively for Westime's clientele. Every member of Westime's affable, multi-lingual staff is dedicated to providing exceptional service—from explaining the specifics of complications, to hand-delivering a watch across the country. In 2012, as part of its Business Excellence Awards, the Beverly Hills Chamber of Commerce recognized Westime for its contribution to the community.

◂ **TOURBILLON SAPPHIRE**
(Bulgari)

JACOB&CO
SWISS MADE
THE BRILLIANT MYSTERY
21.58 ct. Invisible Set Baguette Diamonds
Mystery Dial Design
38mm 18K White Gold Case
JACOB&CO
jacobandco.com
+1.212.719.5887
48 East 57th Street
New York, NY

GIOIELLO

de GRISOGONO EXCLUSIVE 1ST MECHANICAL TOURBILLON

LIMITED EDITION OF 10 PIECES

WATCH IN WHITE GOLD WITH BLACK PVD COATING
SET WITH 3.52 CARATS OF WHITE DIAMONDS

WATCH IN WHITE GOLD WITH BROWN PVD COATING
SET WITH 8.77 CARATS OF BROWN DIAMONDS

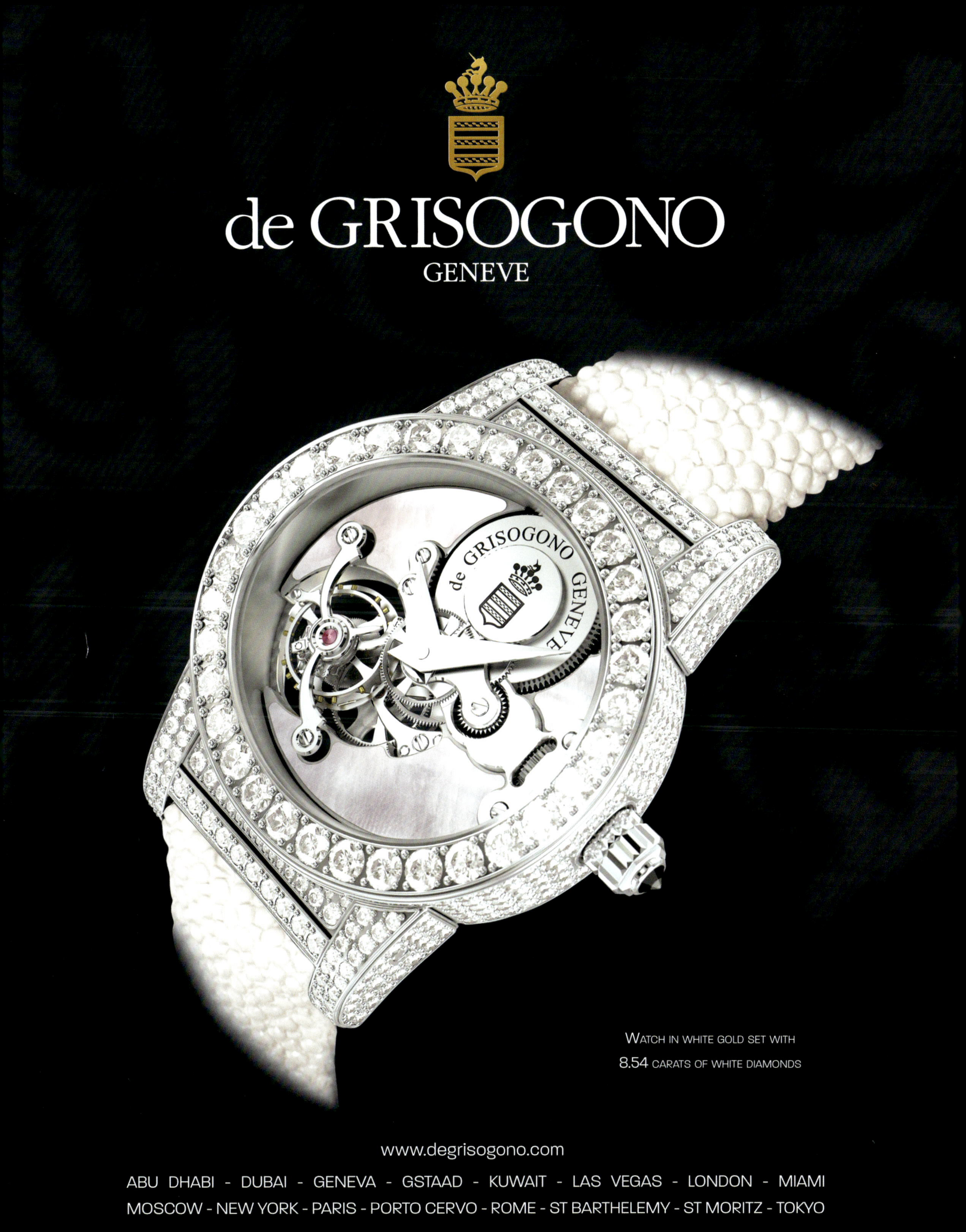
de GRISOGONO
GENEVE
de GRISOGONO GENEVE
WATCH IN WHITE GOLD SET WITH
8.54 CARATS OF WHITE DIAMONDS
www.degrisogono.com
ABU DHABI - DUBAI - GENEVA - GSTAAD - KUWAIT - LAS VEGAS - LONDON - MIAMI
MOSCOW - NEW YORK - PARIS - PORTO CERVO - ROME - ST BARTHELEMY - ST MORITZ - TOKYO

Web Site Directory

A. LANGE & SÖHNE	www.alange-soehne.com
AUDEMARS PIGUET	www.audemarspiguet.com
BELL & ROSS	www.bellross.com
BLANCPAIN	www.blancpain.com
BREGUET	www.breguet.com
CHANEL	www.chanel.com
CHRISTOPHE CLARET	www.claret.ch
DE BETHUNE	www.debethune.ch
DE GRISOGONO	www.degrisogono.com
DIOR HORLOGERIE	www.diorhorlogerie.com
F.P. JOURNE	www.fpjourne.com
FRANCK MULLER	www.franckmullerusa.com
FRÉDÉRIQUE CONSTANT SA	www.frederique-constant.com
GLASHÜTTE ORIGINAL	www.glashuette-original.com
GIRARD-PERREGAUX	www.girard-perregaux.com
GUY ELLIA	www.guyellia.com
H. MOSER	www.h-moser.com
HARRY WINSTON	www.harrywinston.com
HAUTLENCE	www.hautlence.com
HUBLOT	www.hublot.com
IWC	www.iwc.com
JACOB & CO.	www.jacobandco.com
JAEGER-LECOULTRE	www.jaeger-lecoultre.com
LONGINES	www.longines.com
PANERAI	www.panerai.com
PARMIGIANI	www.parmigiani.ch
PATEK PHILIPPE	www.patek.com
PIAGET	www.piaget.com
RICHARD MILLE	www.richardmille.com
RJ-ROMAIN JEROME	www.romainjerome.ch
ROGER DUBUIS	www.rogerdubuis.com
TAG HEUER	www.tagheuer.com
VACHERON CONSTANTIN	www.vacheron-constantin.com
ZENITH	www.zenith-watches.com

RELATED SITES

BASELWORLD	www.baselworld.com
SIHH	www.sihh.ch

AUCTION HOUSES

CHRISTIE'S	www.christies.com
SOTHEBY'S	www.sothebys.com

Bell & Ross
8
55
5
50
10
45
15
40
20
35
30
25
BAS
HAUT

Index

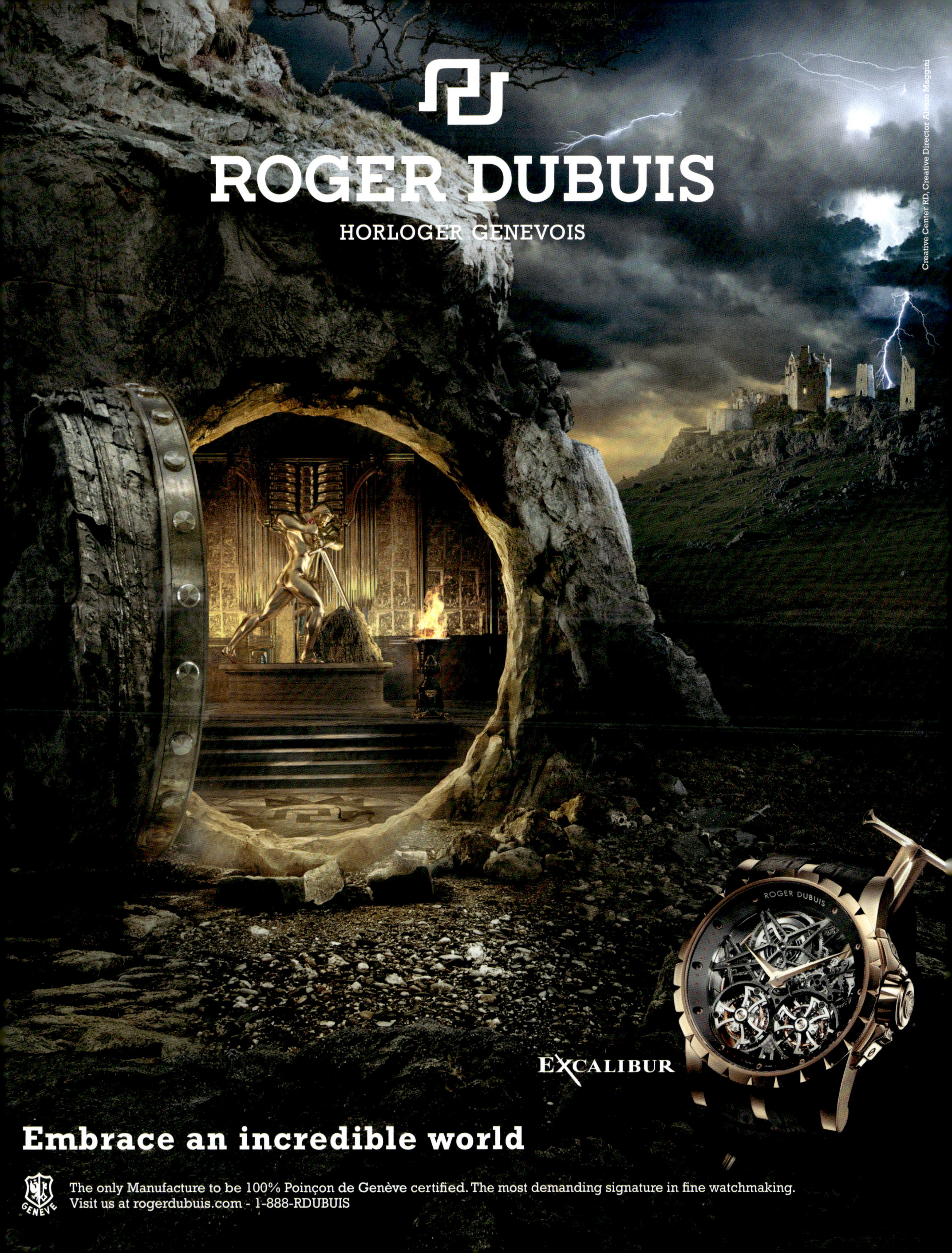
ROGER DUBUIS
HORLOGER GENEVOIS
Creative Center RD, Creative Director Alvaro Maggini
ROGER DUBUIS
EXCALIBUR
Embrace an incredible world
GENÈVE
The only Manufacture to be 100% Poinçon de Genève certified. The most demanding signature in fine watchmaking.
Visit us at rogerdubuis.com - 1-888-RDUBUIS

DB 25
PERPETUAL CALENDAR
CONTEMPORARY EXPRESSION OF WATCHMAKING ART
PERPETUAL CALENDAR
SPHERICAL MOON PHASE
TITANIUM/PLATINUM BALANCE WHEEL
TRIPLE PARE-CHUTE SHOCK-ABSORBING SYSTEM
DE BETHUNE
L'ART HORLOGER AU XXIe SIÈCLE

Index

H. Moser & Cie.

MOSER PERPETUAL 1, ref. 341.501-004. 18K rose gold. Hand-wound movement cal. HMC 341.501. 10 days power reserve. Back and forth date adjustment via the crown. Instantaneous date change, even from Feb. 28th to March 1st. Winner of the prestigious «Grand Prix d'Horlogerie de Genève» 2006 category «Montre Compliquée»

HM & Cie

www.h-moser.com

Elegance is an attitude
Simon Baker

LONGINES®
Conquest Classic

Elegance is an attitude
Kate Winslet

LONGINES®
Conquest Classic
www.longines.com

Index

10000
1/100
7 200 000 BEATS PER HOUR
SWISS MADE
5/10000th (0.00XX)
GPHG
GRAND PRIX D'HORLOGERIE DE GENÈVE
2012
"Aiguille d'Or" Grand Prix

Index

FREDERIQUE CONSTANT
GENEVE
FREDERIQUE CONSTANT
GENEVE
Live your passion
with Classics Manufacture
Innovation and uncompromising quality are the hallmarks of Frédérique Constant. Driven by an unparalleled passion for precision and craftsmanship, our watchmakers manufacture Geneva timepieces of contemporary, classic design and exceptional value.
Contact. 1.1-877-61-WATCH . www.frederique-constant.com

Index

ZENITH
SWISS WATCH MANUFACTURE
SINCE 1865
EL PRIMERO
CHRONOMASTER 1969
www.zenith-watches.com
Life is in the movement

Minute Repeaters

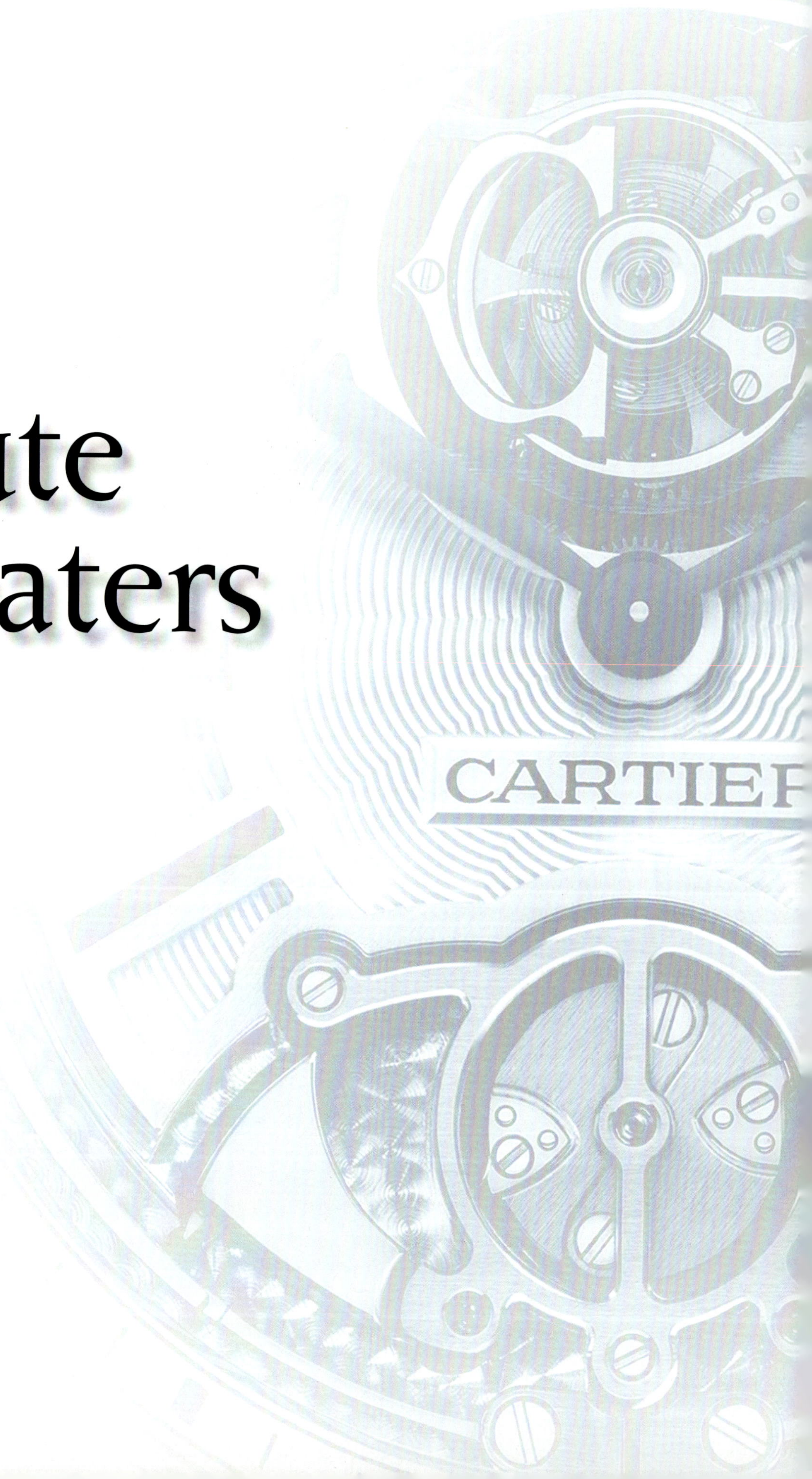

CARTIER
SWISS MADE

Marc A. Hayek

President of Breguet, Blancpain and Jaquet Droz

"The three brands are really very different"

MARC-ALEXANDRE HAYEK, grandson of Nicolas Hayek and son of Nayla Hayek—current Chair of the Swatch Group Board of Directors—is at the head of the Breguet, Blancpain and Jaquet Droz brands. He explains the distinctive characteristics of this threesome that represents the upper segment of the world's no. 1 watchmaking group.

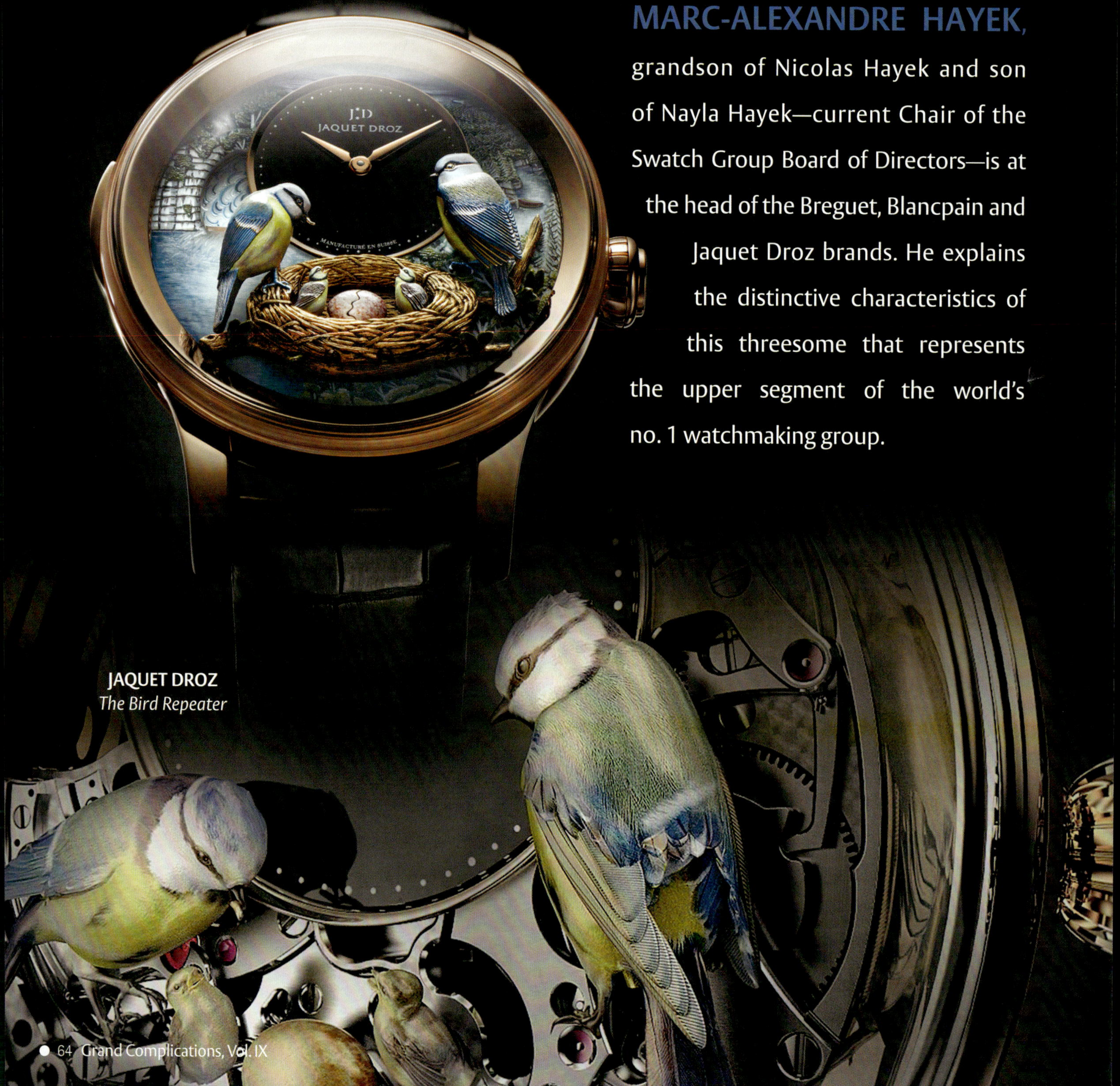

JAQUET DROZ
The Bird Repeater

Breguet, Blancpain and Jaquet Droz watches constitute the top of the range at Swatch Group. What are the differences between them and what is their specific identity?

Breguet represents the essence of research and technological innovation—and has done so ever since the brand was founded by Abraham-Louis Breguet. We are currently striving to pursue this tradition and build on this heritage, obviously with the modern means now available to us.

Is it an avant-garde brand?

Yes, albeit with all the respect we have always shown for tradition. Breguet has always been avant-garde.

What about Blancpain?

This is also a brand with a longstanding tradition, although it has experienced several development stages and displayed various faces. In the 1950s, for example, Blancpain offered purely sporty watches; and in the 1980s, far more classic and traditional ones. It is this diversity that makes the brand interesting: youthful, strongly movement-oriented and endowed with a pure, uncluttered style that nonetheless carries the unmistakable Blancpain signature.

And Jaquet Droz?

This might be seen as the philosophical brand of the Swatch Group's upper segment. Witness one of our latest models, The Bird Repeater, a minute repeater enlivened by songbirds on the dial: a truly magnificent creation! I would never have adopted this approach for either Breguet or Blancpain. That is why artistic crafts are so important at Jaquet Droz. They represent its creative force, which lies in its ability to constantly amaze. These three-dimensional birds are reminiscent of Pierre Jaquet-Droz's beloved automatons. The brand is thus pursuing a more philosophical and poetic path than that of the more traditional brands.

Have the differences you point out always existed, or are you accentuating them because it is essential to distinguish between these three brands? In other words, are Blancpain and Breguet more different today than they were five years ago?

I don't think they are, since the three brands have fortunately always been historically very different. If that were not so, I would have found it very tough to handle them simultaneously. Even if one looks back a century or more, the three firms were already very different from each other—perhaps even more so than they are today. Naturally, both Breguet and Blancpain make tourbillons, and they are competitors because they belong to the same segment. But quite obviously, an historical model such as a Breguet aviator watch has a history totally different from that of the Blancpain Fifty Fathoms. I've never really had to look for the differences between the two brands, and I have never had to intervene in the creation process of a watch at one of them because its style was too similar to the other…

BREGUET
Tradition Breguet 7047
Tourbillon Fusée
Rose Gold

BREGUET
Type XXI Titanium

Which are the models, or more broadly speaking, the horological complications most representative of each of these three brands?

I hardly need to remind people that Abraham-Louis Breguet is the inventor of the tourbillon! The model most emblematic of Breguet is in my opinion the Tourbillon Fusée, which is part of the Tradition collection. It is a concentrated blend of high technology and extreme tradition.

As far as Blancpain is concerned, I would say the Fifty Fathoms. This is the model I can relate to most closely. At the same time, the complication that symbolizes Blancpain is the moonphase. As I said, Blancpain is a multi-faceted brand, which makes it hard to narrow it down to just one model.

Finally, for Jaquet Droz, I really think that the most emblematic model is the Grande Seconde. Nonetheless, the one that best represents the brand, and the challenges it intends to rise to in the future, is in my view The Bird Repeater. Strongly akin to the automaton, this model embodies the Pierre Jaquet-Droz spirit more than any traditional watch.

BLANCPAIN
Fifty Fathoms

Does this model herald what Jaquet Droz will be about in the coming years?

It represents a step in this direction, and there will be others. But this watch that is exceptional in all ways will never replace the traditional Jaquet Droz collection—which is why I contend that the Grande Seconde represents the brand's work, while The Bird Repeater reflects the spirit of Pierre Jaquet-Droz.

Can we expect another new model of this type in 2013?

Yes, at this year's BaselWorld we will be presenting a new Jaquet Droz watch inspired by the founder's work.

BLANCPAIN
Traditional Chinese Calendar

From a far more global standpoint, what do you see as the main issues facing watchmaking in the coming years?

Quality, most definitely; investment in R&D and in production facilities of course; and finally the Swiss Made label. In the latter field, even though we are scarcely affected, as our production is 100% Swiss Made, we believe the Swiss industry as a whole has the responsibility of strengthening this label and not debasing it. Swiss watchmaking must not rest on its laurels, but on the contrary do all in its power to defend the high standards of Swiss quality. If we start allowing ourselves to take shortcuts, it might work for a few years, but would doubtless end in disaster over the long term. Another challenge for watchmakers lies in ensuring the quality of worldwide distribution. This is an essential factor, because service must always and without exception be on a par with our products. That is the least a customer can expect from us.

JAQUET DROZ
Grande Seconde SW Red Gold

A close-up look at minute

from the verb montrer, meaning "to show" and initially used in the 15th and 16th centuries for the dial alone (including on a clock) before coming to refer to a portable timepiece. Nonetheless, from the very start of mechanical horology, there were instruments providing an audible indication of the time, and church bells rang through town and country well before watches took their place on everyday people's wrists.

repeaters and sonneries

Over the centuries, striking watches—whether of the "in passing" or automatically sounding type, or activated on demand (the famous "repeaters")—have lost their practical usefulness, but never their prestige. On the contrary, they have established themselves more than ever as one of the pinnacle achievements of the grand art of watchmaking, and as the exclusive preserve of a handful of authentic manufactures and gifted artisans.

In recent years, striking watches and particularly minute repeaters have been enjoying a remarkable ascent. Watchmakers are making the most of the latest advances in the fields of micromechanics, acoustics, computer technology and the science of alloys to offer timepieces attaining ever-greater heights of musical perfection.

Welcome to the enchanting world of minute repeaters and other striking watches!

Striking systems

were born at the same time as mechanical horology—which is hardly surprising in Medieval Europe, where the cadence of daily activities was generally marked out by the chiming of bells that were struck by hand.

From the emergence of the first monumental clocks, in around 1300, some were equipped with mechanisms serving to "sound out" the time at a distance and by night, whereas dials and hour hands did not appear until a century later.

In the 14th century, the monumental clock with automatic chime, a symbol of prestige, power and progress, was part of the "adornment" of any self-respecting town—whether it sat enthroned on a religious edifice, a palace, a belfry or a gateway. It was often equipped with jacks, meaning automaton figures that mimed the gesture of striking the hours.

Wall clocks from 1350 onward and subsequently table clocks were also often equipped with chiming systems or striking bells, and often enhanced by automatons, such as a lion opening its mouth and beating its wings.

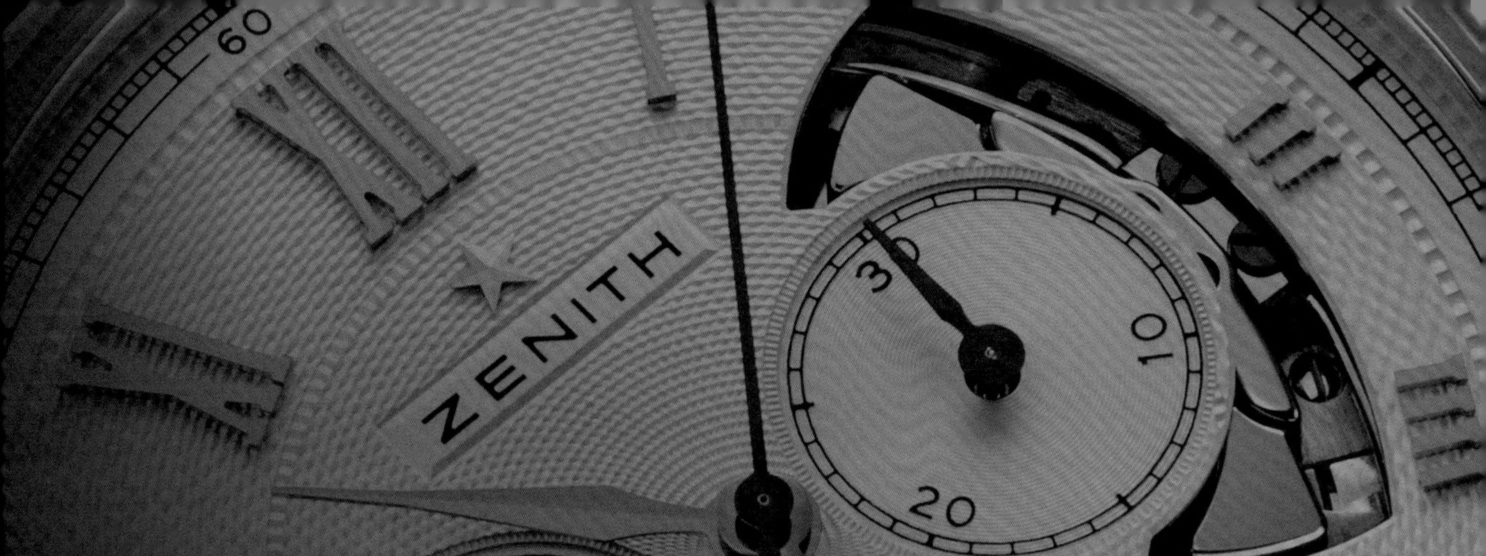

As soon as pocket watches appeared, they were also often equipped with audible mechanisms. In 1488, a letter from an ambassador to the Duke of Milan signals the existence of three watches, including two "with small bells." In the 14th century, records show numerous examples of this type of watch striking the hours "in passing," meaning automatically, just like steeple clocks. The bell (with or without a pendulum) serving to produce the sound was fixed to the back cover, which was openworked to allow the sound to pass through, an approach that gave rise to various decorative fancies. The fashion for striking watches continued throughout the 17th century. In 1642, in his comedy *Le Menteur*, the French playwright Corneille depicts a hidden character betrayed by the chiming of his timepiece!

Moreover, a new type of striking mechanism was soon to make its appearance: the famous "repeater" watches serving to sound the time at will. The first was the "quarter repeater," striking the hours and quarter hours. It was invented in around 1675 by several English watchmakers—Edward Barlow, Thomas Tompion and Daniel Quare. The latter is most generally credited with this achievement, having been granted a patent for the invention in 1687.

The minute repeater, which was long believed to have been invented in around 1750 by another English watchmaker, Thomas Mudge, actually first appeared between 1700 and 1710, notably in southern Germany. Striking the hours, quarter-hours and minutes on demand, it asserted itself as an extremely practical instrument for telling the time at night, as well as serving, like many other timepieces, as a symbol of prestige displaying its owner's power and wealth.

Repeater watches were much in favor during the 18th century. In Geneva, the regulations of the Watchmakers' Corporation stipulated that anyone wishing to join must demonstrate his ability to make a quarter repeater movement—a sure sign of the extreme sophistication of this mechanism. Various "programming" variations emerged, including ten-minute, half-quarter and five-minute repeaters. A number of repeater mechanisms, along with an alarm function, were also found on large "carriage watches" that were eminently useful to travelers. All these striking timepieces were equipped at the time with hammers striking a small bell. For more discreet use, some watches were equipped with hammers delicately striking the case, or with sharp pins protruding from the case to prick the finger—enabling the wearer to tell the time at will while respecting the rules of courtesy, since checking the time in public could be considered impolite.

In the 1780s, the famous Abraham-Louis Breguet had the idea of replacing bells by gongs, meaning steel spring blades, initially placed across the back plate, and then wrapped around the movement inside the case. This crucial invention served both to achieve a clearer, purer sound and to reduce the thickness of the case. Breguet himself made a certain number of repeater watches, including quarter repeaters, and a minute repeater mechanism was one of the many horological complications featured in the famous Marie-Antoinette wristwatch that Breguet began making in 1783 (see the chapter on Multi-Complications).

Fancy watches reigned supreme around the 18th and 19th centuries, which witnessed the emergence of countless creations combining audible mechanisms with automaton systems enlivened by small figures.

From 1845 onwards, the invention of so-called "safety" matches, followed by the emergence of hands and dial with a luminescent coating, deprived repeating watches of their initial reason for being. Nonetheless, thanks to their technical sophistication and the magic of their sound, striking watches—and particularly minute repeaters—earned a definitive place in the horological hall of fame, right up among the rarest and most sought-after complications.

In the 20th century, by then miniaturized to wristwatch size—a change that involved a number of technical and acoustic challenges—watches with an audible indication of time asserted themselves as one of the greatest demonstrations of watchmaking know-how, and many observers indeed came to view them as the supreme complication.

Recent years have seen a constantly growing spate of minute repeater wristwatches—micromechanical masterpieces generally carrying six-figure price tags. To optimize sound and enhance acoustic performances, brands invent new striking systems and explore all the potential of high-tech metal alloys, where necessary using cutting-edge software to control the various parameters of sound.

After the tourbillon tidal wave, we are now seeing a progressive rise of striking watches, and particularly of the minute repeater, which has become the supreme symbol of excellence. As a token of this prestige, minute repeaters have recently been popping up in the collections of brands normally not associated with expertise in complications—generally thanks to the backstage assistance of certain discreet suppliers or master-watchmakers.

As for the Grande Sonnerie, or grand strike complication, which first appeared in wristwatch form in 1992 thanks to Philippe Dufour, it remains an even rarer and more exclusive pleasure reserved for super-complicated watches and for a few exceptional timepieces.

HOW DOES IT WORK?

Operation of a minute repeater

A minute repeater mechanism all by itself calls for over one hundred sometimes microscopic parts, which must all be machined with extreme accuracy. Its highly complex operation can hardly be described in detail in just a few lines. To simplify matters, one may start by pointing out several main parts or groups of parts.

ACTIVATION SYSTEM

This generally consists of a bolt or slide-piece that the user shifts along the caseband, and sometimes of a pushbutton or a pivoting bevel. Activating this system both reloads the mainspring of the audible mechanism and triggers the striking mechanism.

SNAILS

The role of the snails is to reproduce below the dial the hour shown by the hands, in such a way that the striking mechanism can keep track at any time of the number of notes to be struck. Minute repeaters are equipped with three snails, each with a specific shape:

- the spiral-shaped hour snail has 12 notches corresponding to the 12 hours of the day and night
- the asymmetrical quarter snail has 4 notches for the 4 quarter-hours
- the propeller-shaped minute snail as 14 notches on each of its 4 blades corresponding to the 14 last minutes of each quarter (the 1st minute corresponds to the quarter-hour as such).

The quarter and minute snails are fixed to the arbor of the minute hand. After each complete rotation (four quarters), the quarter snail moves the hour snail 30° (one hour) forwards.

HOUR, QUARTER-HOUR AND MINUTE PIECES

These three key components featuring a highly complex shape fulfill a two-fold role by:

- determining the angular position of the snails thanks to "feeler-spindles" positioned at one end
- setting off the right number of blows of the hammer thanks to the "rack" positioned at the other end.

The hour rack has 12 teeth serving to sound up to 12 notes; the quarter rack has two groups of three teeth serving to sound up to three double high-low notes; and the minute rack has 14 teeth serving to sound up to 14 notes.

HAMMER AND GONGS

The striking mechanism itself is composed of hammers striking on gongs. The gongs are thin and generally round steel blades, wrapped around the movement. They are fixed to one of its ends, while the other is free. A minute repeater has two gongs—one high-pitched and the other low-pitched. The hammers take the form of small steel levers pivoting on one axis and held in their resting position by a spring. Once lifted by a gathering-pallet, the hammer briefly strikes the corresponding gong in order to make it vibrate.

GOVERNOR

To guarantee that the blows are struck at regular intervals, minute repeater watches are equipped with a governor (also known as a regulator) that controls the striking speed of the hammers. There are various governor systems, including some using centrifugal force.

An example of a watch sounding 11:37

To explain how the entire mechanism works, the simplest way is to take a concrete example.

11 o'clock… Activating the repeater slide-piece charges the striking mechanism mainspring. It also sets into motion the hour piece, which pivots until its feeler-spindle hits the 11th notch of the hour snail (i.e. 11 o'clock). During this pivoting process, the first 11 teeth of the hour rack successively activate the hour hammer. The watch sounds 11 notes on the low-pitched gong.

…30… After the hours have sounded, the winding lever releases the quarter piece. Its feeler-spindle hits the 2nd notch of the quarter snail (30 minutes = 2 quarters). During this pivoting process, the 2 first teeth of each of the 2 groups of teeth on the quarter rack alternately activate the 2 hammers. The watch sounds 2 double high-low notes.

…7 minutes After the double blows, the quarter rack releases the minute piece. Its feeler-spindle hits the 7th notch on the 2nd "blade" of the minute snail. During this pivoting process, the 7 first teeth of the minute rack successively activate the minute hammer. The watch sounds 7 high notes.

The watch has struck 11 hours, 2 quarters (30 minutes) and 7 minutes: 11:37.

Making a minute repeater

Making a minute repeater is a lengthy, meticulous and complex process that requires a combination of peerless technical skills and an excellent musical ear. It is reserved for an elite group of artisans with solid experience in this domain.

The master-watchmaker must in particular shape the gongs by hand so as to ensure they follow the shape of the movement, and then fit them inside the confined space of the case without them touching each other, while positioning them extremely accurately in relation to the hammers. He must also adjust the length and the thickness of each gong in order to achieve the right note—just like a musician tuning an instrument. The process of securing the gongs, testing the sound, dismantling, adjusting and reassembling the mechanism may take several hundreds of hours.

Adjusting the hammer springs also plays an essential role: the weight of the hammer must be perfectly adjusted so as to achieve the right sound, neither too loud nor too soft.

The watches are then subjected to extremely advanced acoustic tests serving to measure the pitch, the intensity and the duration of the sound produced by each of the gongs.

Grande Sonnerie watches

Grande Sonnerie (grand strike) models, which sound the hours and quarters in passing and repeat the hour every quarter-hour, are regarded as the most sophisticated of watches providing an audible indication of the time. They call for a number of parts far superior to the minute repeater (with which it is often associated). One of its main challenges lies in mastering energy. Whereas a minute repeater is reloaded each time the repeater slide-piece is activated, a Grande Sonnerie must at all times have sufficient energy to strike the desired number of notes (96 notes in 24 hours!) without running down after only a few hours, and while consistently producing a sound of the same quality. To achieve this, it is generally equipped with two mainsprings: one for the timekeeping gear trains, and the other for the striking mechanism.

The basics of striking watches

REPEATING WATCHES

Watches equipped with a mechanism enabling them to sound the time on demand when a slide-piece or pushbutton is activated. "Quarter repeaters" successively indicate the number of hours and then the quarter-hours shown on the dial. There are also "half-quarter repeaters" and "ten-minute repeaters," as well as "five-minute repeaters" and "minute repeaters."

MINUTE REPEATERS

The most sophisticated type of repeater watch, the "minute repeater," sounds the time on demand to the nearest minute. These watches are generally equipped with two gongs: one low-pitched for the hour, and the other high-pitched for the minutes, with the quarters marked by a succession of high-low notes. Activating the minute repeater at 8:47 makes it sound 8 low notes, three double high-low notes and two high notes.

GRANDE AND PETITE SONNERIES (GRAND AND SMALL STRIKES)

Grande and Petite Sonnerie watches automatically sound the hours and quarters in passing, like a church steeple clock. Grande Sonnerie models repeat the hour before each quarter; Petite Sonnerie models merely sound the hours or quarters (without repeating the hour at each quarter). A "silent" mode that can be activated by a slide-piece or a pushbutton serves to deactivate the striking mechanism at will.

GONGS

Thin steel blades struck by the hammers. The gongs are wrapped inside the case all around the movement and fixed at one end. Depending on their length, they produce a lower or higher-pitched sound. The type of alloy plays a crucial role in terms of the quality of the sound.

"CATHEDRAL" GONGS

"Cathedral" gongs are gongs that are wrapped more than once—generally almost twice—around the movement and thereby generate a richer and deeper sound.

CARILLON CHIME

A carillon chime is a striking mechanism comprising at least three gongs.

WESTMINSTER CHIME

So-called Westminster chimes sound each quarter by successively striking four different gongs. Contrary to what their name might imply, they do not necessarily play the tune of the famous Big Ben—the clock on the British Parliament—originally composed by Handel.

Minute Repeaters: Innovation at Work

In recent years, watchmakers seem determined to revive interest in minute repeater models by exploring new technical paths, with changes to construction or affixation of the chimes, alloys used for their fabrication, transmission of the sound, ways of activating the strike or the increasing water resistance of these ultra-sophisticated watches.

Breguet
Classique Grande Complication Minute Repeater 7637
Beneath its traditional appearance, the Classique Grande Complication Minute Repeater from Breguet conceals an entirely reinvented striking mechanism that uses new materials and innovative positions for the chimes, chime-holders and hammers.

IWC
Portuguese Minute Repeater
IWC has opted for a sober, classical aesthetic on its Portuguese Minute Repeater, equipped with a large case in platinum or rose gold and a manual-winding pocket watch movement.

Cartier, *Rotonde Minute Repeater*

On its Rotonde Minute Repeater watch, Cartier offers a new way of securing the gongs aimed at achieving perfect acoustic synergy between the case and the movement. The square-section gongs enable the hammer to regularly strike the same spot.

Patek Philippe, *Ref. 5074*

In its Ref. 5074, Patek Philippe—which has a collection comprising around 13 minute repeater wristwatches—has combined this complication with a perpetual calendar. Its "cathedral" gongs feature a particularly lengthy resonance time.

Jaeger-LeCoultre, *Master Grande Tradition Minute Repeater*

Equipped with a silicon escapement, the Master Grande Tradition Minute Repeater from Jaeger-LeCoultre benefits from an unusual system for transmitting sound, in which square-section gongs are welded directly to the sapphire crystal—a system that was patented by the manufacture.

Girard-Perregaux, *1966 Minute Repeater*

To endow its 1966 Minute Repeater with an optimal sound, Girard-Perregaux has reworked the construction of the watch, notably by seeking the best possible relationship between the diameter of the caliber and the inner diameter of the case, and by curving the back to increase the volume of air between the movement and the case.

Christophe Claret, *Adagio*

On this minute repeater model from Christophe Claret, Adagio, the cathedral chimes of the striking mechanism are equipped with a patented device that prevents the chimes from hitting each other while they vibrate under the hammers.

Bulgari, *Daniel Roth, Carillon Tourbillon*

The Daniel Roth Carillon Tourbillon by Bulgari is distinguished by its striking mechanism with three hammers and three gongs, which are extremely rare on this type of model. It sounds the hours on a C note, the quarters with an E-D-C sequence, and the minutes on an E note.

Graff Luxury Watches

MasterGraff Diamond Minute Repeater Tourbillon

Equipped with an original case inspired by the shape of a diamond—an exclusive signature feature of Graff watches—the MasterGraff Diamond Minute Repeater Tourbillon incorporates two of the most sophisticated complications. The flying tourbillon is visible through an opening at 6 o'clock.

Manufacture Royale, *Opéra*

The first creation from the hyper-exclusive brand Manufacture Royale, the Opéra tourbillon and minute repeater model stands out for its case, which resembles a metallic accordion bellows and acts as a resonance chamber for the gongs.

Audemars Piguet

Jules Audemars Minute Repeater Tourbillon Chronograph

On its Jules Audemars Minute Repeater Tourbillon Chronograph, Audemars Piguet resolved to develop a new striking system, with new hammers as well as two gongs crafted all of a piece for a more consistent tone.

Corum

Admiral's Cup Minute Repeater 46 Acoustica

Developed in collaboration with La Fabrique du Temps, the Admiral's Cup Minute Repeater 46 Acoustica by Corum stands out for its four gongs and four hammers, operating in pairs to create harmonies. An eight-piece worldwide limited edition.

Corum

Admiral's Cup 45 Minute Repeater

Doing away with the traditional slide-piece on the side of the case, Corum has devised for its Admiral's Cup 45 Minute Repeater an ingenious mobile bezel serving to load the minute repeater by a simple 27° clockwise rotation. This option notably ensures water-resistance to 30m.

Roger Dubuis, *Excalibur Minute Repeater*

A new kind of function-starting device equips Roger Dubuis's Excalibur Minute Repeater: the wearer activates the strike with a turn of the rotating bezel.

Jaeger-LeCoultre
Reverso Répétition Minutes à Rideau

Unveiled in 2011, Jaeger-LeCoultre's Reverso Répétition Minutes à Rideau invented a new sensation with a sliding shutter system that hides one of the faces of the case—while activating the strike mechanism.

Cvstos, *Challenge RMT-S*

The Challenge RMT-S from the Genevan brand Cvstos, with anti-shock tourbillon carriage, combines a minute repeater with water resistance to 100m—in a distinctly futuristic manner.

Blancpain
Léman Minute Repeater Aqua Lung

Blancpain has opted to tackle one of the trickiest issues with minute repeater watches—the Achilles heel in their water resistance represented by the slide-piece—with its Léman Minute Repeater Aqua Lung, water resistant to 100m.

Vacheron Constantin
Ateliers Cabinotiers Philosophia

A one-of-a-kind model made to order by Vacheron Constantin, the Philosophia has only one central hand, that of the hours, providing a 24-hour reading of the time. To gain a more precise idea of the time at any given moment, the user need only activate the minute repeater slide-piece.

VACHERON CONSTANTIN
GENÈVE
PHILOSOPHIA

Minute Repeaters: Open-Hearted Mechanisims

On classical minute repeater watches, nothing signals the presence of this sophisticated complication—besides the presence of a small slide-piece or bolt on the side of the case. Following a general trend in haute horology, many watchmakers today prefer to reveal the mechanism by openworking the dial, or forgoing it altogether.

Speake-Marin

Renaissance Tourbillon Minute Repeater

Created by the independent master-watchmakers Peter Speake-Marin, the Renaissance Tourbillon Minute Repeater is fitted with an openworked dial revealing the intricate inner workings of the mechanism, as well as the impressive tourbillon at 4:30.

Zenith, *Academy Minute Repeater*

In the new Academy Minute Repeater, Zenith has combined its famous automatic El Primero chronograph movement with a striking mechanism visible through a dial opening. This model is available in a 25-piece pink-gold limited edition.

Christophe Claret, *Soprano*

With its new Soprano model, Christophe Claret pursues its quest for technical and aesthetic excellence by associating a tourbillon with a minute repeater featuring a Westminster chime comprising four patented gongs and four hammers. This impressive feat is beautifully highlighted by stepped bridges inspired by the Charles X style.

Audemars Piguet
Millenary Minute Repeater

Housed in an oval brushed titanium case, the Millenary Minute Repeater from Audemars Piguet stands out for its three-dimensional architecture that reveals the new Audemars Piguet-patented escapement, a double balance-spring and the hammers and gongs of the striking mechanism.

Le Brassus Minute Repeater Carrousel

Le Brassus Minute Repeater Carrousel from Blancpain associates a flying carrousel (see chapter on Tourbillons) with a minute repeater whose cathedral chimes are affixed to the interior side of the case, a configuration that provides a sound of peerless volume and clarity.

Hublot

Minute Repeater Cathedral Tourbillon and Column Wheel Chronograph

Equipped with a carbon fiber case and bridges, the Minute Repeater Cathedral Tourbillon and Column Wheel Chronograph by Hublot is also distinguished by its "cathedral" gongs that are twice as long as standard gongs. The three complications are visible from the dial side.

Claude Meylan
Répétition 5 (Ref. 8731)
Launched in 2012, the Répétition 5 (Ref. 8731) by Claude Meylan features a movement entirely openworked and engraved by hand. Pressing the pushbutton at 8 o'clock makes the watch sound the number of hours, followed by the number of five-minute intervals elapsed since the previous hour.

Seiko, *Credor Spring Drive Sonnerie*
The Japanese brand Seiko has revived the antique hour repeater mechanism in a high-end model named Credo Spring Drive Sonnerie. The mechanism strikes in three modes: the number of hours on each hour; three strikes every three hours (3, 6, 9 and 12 o'clock); and silent mode.

deLaCour, *Birépétition Minutes*
Equipped with a Christophe Claret movement, the Birépétition Minutes from deLaCour strikes the hours, quarter-hours and minutes for either the local time or a second time zone.

Franc Vila, *FVa35 SuperSonico 5-Minute Repeater*
To improve the acoustic qualities of the striking mechanism, Franc Vila has equipped its FVa35 SuperSonico 5-Minute Repeater with a titanium and red-gold case and with Lightnium bridges—an aluminum and lithium alloy used in aeronautics.

Louis Vuitton
Tambour Répétition Minutes
On the Tambour Répétition Minutes from Louis Vuitton, with a dual time zone, the minute repeater sounds the home time, not local time—a pleasant way to feel connected with close friends and family.

Grand and Small Strikes

Like a church bell on the wrist, grand and small strike systems, also called grande et petite sonnerie, mark the hours and quarter-hours in passing (with grand strikes repeating the hours each time). This extremely rare function is the prerogative of ultra-complicated wristwatches (see chapter on Multi-complications). We also find it in some highly sophisticated watches where it most often joins the minute repeater—sometimes in chiming the famous melody of Big Ben.

Patek Philippe, *Star Caliber 2000*

Among its 21 complications, the Star Caliber 2000 pocket watch by Patek Philippe has a mechanism striking in passing and a minute repeater with Westminster chime on five gongs faithfully reproducing the tune rung by the Big Ben clock on the Houses of Parliament in London.

Bulgari, *Grande Sonnerie Perpetual Calendar*

The Bulgari Grande Sonnerie Perpetual Calendar has a chime playing the Westminister tune on four notes. It is also equipped with two barrels, of which one is exclusively dedicated to the striking mechanism and guarantees a 24-hour power reserve in grande sonnerie mode.

Franck Muller, *Aeternitas Mega 4*

Among its fifteen or so complications, the Aeternitas Mega 4 by Franck Muller features a grande and petite sonnerie as well as a minute repeater with four-gong Westminster chime—all housed within an elegant tonneau-shaped case.

Jaeger-LeCoultre

Hybris Mechanica à Grande Sonnerie

The Hybris Mechanica à Grande Sonnerie is part of the Hybris Mechanica trilogy of ultra-complicated watches presented by Jaeger-LeCoultre (see chapter on Grand Complications). Its movement, comprising over 1,300 parts, incorporates a grand and small strike as well as a minute repeater, along with a Westminster chime that sounds a tune that lasts much longer than normal.

Philippe Dufour, *Grande Sonnerie*

Launched in 1992, the Grande Sonnerie from Philippe Dufour—one of the preeminent members of the Académie Horlogère des Créateurs Indépendants—was the first minute repeater watch with grand and small strike.

F. P. Journe
Sonnerie Souveraine

Possessing grand strike and minute repeater capabilities, the Sonnerie Souveraine is distinguished by its extreme functionality and security assured by ten patented devices. F.P. Journe actually designed it "so an eight-year-old child could use it without damaging it"—which required a high degree of technical sophistication in the conception and design.

A. Lange & Söhne
Zeitwerk Striking Time

The Zeitwerk Striking Time from A. Lange & Söhne strikes each quarter-hour in passing with a high note, then the hour with a low one—thus reproducing in slow motion the famous tune from the beginning of Beethoven's Fifth Symphony.

Chopard, *L.U.C Strike One*

On its L.U.C Strike One, Chopard presents a mechanism striking each hour with a single note (hence its name), while an opening in the dial serves to reveal a small hammer at work.

Audemars Piguet, *Grande Sonnerie Carillon*

Presented at the end of the 1990s, the Grande Sonnerie Carillon by Audemars Piguet is distinguished by its dynamograph system indicating the real torque delivered by the barrel. It has a power reserve of 20 hours in Grande Sonnerie mode and 60 hours in Petite Sonnerie mode.

Feminine Melodies

In a still tentative trend, we have recently seen the first watches with minute repeater or "in passing" strike mechanisms created especially for women.

Breguet, *Reine de Naples Sonnerie au passage 8978*

To celebrate the 200th anniversary of the wristwatch created for Caroline Murat, Queen of Naples, Breguet presented the Reine de Naples Sonnerie au Passage 8978, with a mechanism that marks each full hour with two strikes repeated three times.

Ellicott, *Lady Tuxedo Midnight*

Ellicott's Lady Tuxedo Midnight model is equipped with a self-winding movement striking the hours and quarter-hours on request. Its sensual case, partially or entirely set with diamonds, is crafted in rose gold, white gold or steel.

Patek Philippe, *Ladies First Minute Repeater Ref. 7000*

First Patek Philippe minute repeater wristwatch exclusively dedicated to women, the Ladies First Minute Repeater Ref. 7000 is distinguished by its ultra-thin self-winding movement and its pure aesthetic, featuring a rose-gold case .

Watches with Automatons

In terms of a grand visual and acoustic spectacle, watches with automatons, which are extremely rare, unquestionably occupy an echelon all to themselves.

Bovet Fleurier
Amadeo Fleurier 44
Minute Repeater Tourbillon
Triple Time Zone Automaton

The Amadeo Fleurier 44 Minute Repeater Tourbillon Triple Time Zone Automaton watch by Bovet Fleurier features a dial featuring an automaton with bells reminiscent of the first jaquemart (jack) mechanisms.

Van Cleef & Arpels, *Lady Arpels Poetic Wish*

Signed Van Cleef & Arpels, the Lady Arpels Poetic Wish combines a five-minute repeater with three automatons—a young girl, a tiny mother-of-pearl cloud and a kite—moving on request to sound the hours and minutes in music.

Jaquet Droz, *The Bird Repeater*

Heir to the famous 18th-century Jaquet Droz automatons, The Bird Repeater wristwatch features a pair of birds on a nest in front of a stream—a three-dimensional tableau that springs to life each time the cathedral-gong minute repeater is activated.

Ulysse Nardin
Minute Repeater Alexander The Great Westminster Carillon

Issued in a 100-piece worldwide edition, the Minute Repeater Alexander The Great Westminster Carillon by Ulysse Nardin is distinguished by its five jacks depicting Alexander the Great in combat and moving in step with the sound of the gongs.

Daniel Roth
Il Giocatore Veneziano

Inspired by a painting by Caravaggio, Il Giocatore Veneziano from Daniel Roth associates a minute repeater mechanism (with cathedral chime) with an automaton that throws two dice with a total of 504 possibilities. The automaton movement operates independently of the minute repeater mechanism, and may thus be activated either at the same time as the striking mechanism, or on demand.

A. LANGE & SÖHNE

LANGE ZEITWERK STRIKING TIME – REF. 145.029

The most recent addition to the Lange Zeitwerk family is endowed with a chiming mechanism that is visible on its dial. The Lange Zeitwerk Striking Time is the first Lange wristwatch with an acoustic signature. It strikes the quarter-hours with high-pitched tones and the full hours at a lower pitch. The chiming mechanism consists of two black polished steel hammers that are integrated in the dial layout on either side of the subsidiary seconds. The manually wound L043.2 caliber is crafted to the most exacting Lange quality standards and largely decorated and assembled by hand. The sonorous timepiece comes in a 44.2mm white-gold case with a black dial or in a limited edition of 100 pieces with platinum cases and rhodium-plated dials.

AUDEMARS PIGUET

JULES AUDEMARS OPEN-WORKED MINUTE REPEATER – REF. 26356PT.OO.D028CR.01

The Jules Audemars Open-worked Minute Repeater with jumping hour and small seconds is powered by Audemars Piguet's in-house, open-worked, hand-wound Manufacture caliber 2907, which is comprised of 420 parts and beats at 21,600 vph. This timepiece has a 950 platinum case and folding clasp, with a sapphire crystal exhibition caseback. It features an open-worked white dial and is secured to the wrist with a blue alligator leather strap. It displays jumping hour through dial aperture, minutes and small seconds, and there is a minimum guaranteed power reserve of 72 hours.

AUDEMARS PIGUET

MILLENARY HAND-WOUND MINUTE REPEATER WITH AP ESCAPEMENT REF. 26371TI.OO.D002CR.01

The Millenary Minute Repeater is a concentrated blend of expertise, technical sophistication and innovative materials. Its titanium oval case provides an atypical setting for a truly magnificent sight featuring offset gold subdials. The AP escapement and the double balance-spring, as well as the hammers and the striking gongs, all become key players in a truly three-dimensional creation. Conceived, developed and produced by Audemars Piguet, the new hand-wound 2910 caliber driving the Millenary Minute Repeater is also distinguished by the atypical construction of the regulating organ. The oval Millenary case in brushed titanium—a material featuring exceptional acoustic qualities—is framed by a polished titanium bezel. The various organs within are barely concealed by the anthracite gray hours and minutes subdial offset at 3:00, and by the small seconds at 7:00. Limited edition of eight pieces.

BELL & ROSS

PW1 REPETITION MINUTES

Bell & Ross assigns new meaning to the pocketwatch, as its PW1 Répétition Minutes resonates with poetic melody. The watch keeps time like the clocks of yesteryear, as the wearer does not look for the hours to pass, but listens to them. Marked by sounds harmoniously emitted through its Argentium case, the PW1 Répétition Minutes has quite the complex configuration, indicating hours and minutes musically on request. To initiate the complication, the wearer presses a button on the left-hand side of the case. Upon activation, a low-pitched sound is emitted for the hours and a series of double low-pitched and high-pitched sounds indicates each increment of five minutes that have passed.

BLANCPAIN

CARROUSEL REPETITION MINUTES LE BRASSUS – REF. 0233-3634-55B

Blancpain's Carrousel Répétition Minutes Le Brassus is powered by the 1736 caliber residing in its 18K red-gold 45mm case, and features a flying one-minute carrousel. A slide lever operates the minute repeater and a cathedral gong sounds the hour. Historically, repeaters have been vulnerable to damage if the owner attempts to change the time while the repeater is engaged and sounding. An innovation addresses this problem; activating the repeater disconnects the mechanism from the crown, thus eliminating the risk of damage if the crown is manipulated. As befits such a complicated movement, the red-gold movement bridges and mainplate are individually hand-engraved. The combination of an openworked dial center and sapphire caseback allows a perfect view of the complex minute repeater mechanism and hand carving of the movement bridges.

CHRISTOPHE CLARET

ADAGIO – REF. MTR.SLB88.023-823

This exceptional watch is equipped with the 455-component Calibre SLB88, created entirely in-house. The Adagio features central hour and minute hands, small seconds at 9:00 in a window or subdial, along with a large date at 6:00 and a dual time zone (hours and minutes) display complete with day/night indication in a subdial at 2:00. In tribute to his manufacture's first caliber introduced over 20 years ago, Christophe Claret equipped this timepiece with a minute repeater, a particularly intricate complication that has become one of the great specialties of the manufacture. Striking the hours, quarters and minutes on demand, the "cathedral gongs" of the striking mechanism feature a patented device preventing them from knocking against each other when they vibrate under the hammers. The patented large date display boasts an unprecedented construction.

CHRISTOPHE CLARET

SOPRANO – REF. MTR.TRD98.010-018

The Soprano features a musically accurate four-note minute repeater striking Westminster Quarters on patented cathedral gongs, a 60-second tourbillon and Charles X style bridges, all with a spectacular dial-free view. The Soprano is a timepiece of contrasts: traditional haute horology with state-of-the-art manufacturing; English Parliament with French King; historic complications with contemporary design; aural indications with visual displays; noble gold with high-tech titanium, and metal components with sapphire elements. With decades of experience developing striking watches, Christophe Claret has not just mastered the mysterious art of minute repeaters, but has brought the genre into the 21st century.

de GRISOGONO

OCCHIO RIPETIZIONE MINUTI S01

The exceptional Occhio Ripetizione Minuti boasts a mechanism very similar to the one found in a reflex camera. Comprising 12 titanium blades, the diaphragm opens to reveal the movement as the mechanism chimes. The watch's sapphire crystal caseback reveals the movement of the three hammers (to strike the hours, quarter-hours, and minutes) on the gongs, as well as the movement's signature decorative finishing. This model in blackened 18K white gold and platinum features 410 white diamonds (1.89 carats) on the bezel.

F.P. JOURNE

SONNERIE SOUVERAINE

Operating a chiming watch has always been risky, as the slightest mistake in operation can damage the movement. To meet the demand for an easy-to-use timepiece, construction of the Sonnerie Souveraine required ten patents on new mechanisms, including the barrel, power reserve indicator, winding and setting systems and chiming gongs. In this grand strike wristwatch, a single mainspring provides enough energy for 24 hours of grand strike (96 full chimes in passing). The chiming functions alone use up almost 60% of the mainspring's energy, and in silent mode without the chime, the movement runs for five days. Constructing this movement has been a permanent quest to maximize mechanical efficiency. The result is a low-tension movement with mechanisms that have to be very finely adjusted to ensure unfailing chimes 35,040 times a year.

GUY ELLIA

REPETITION MINUTE ZEPHYR

The ambassador of Guy Ellia's complicated timekeeper, the Répétition Minute Zephyr with its complications and intricate manufacturing process, is considered a particularly exceptional piece. The Caliber GEC 88 was created by Swiss manufacture Christophe Claret and features a power reserve indicator, minute repeater function and five time zones with day/night indication. The movement beats at 18,000 vph in a solid, convex, high-resonance sapphire crystal case of impressive proportions (53.6x43.7x14.8mm). Mounted on an alligator strap with a folding buckle, this model is also available in pink gold or titanium.

HARRY WINSTON

HARRY WINSTON MIDNIGHT MINUTE REPEATER – REF. 450/MMMR42RL.K

Harry Winston celebrates one of the most complex grand complications with this limited edition minute repeater. Available in 40 pieces per color, this unique model has an openworked dial, which allows visibility of the minute repeater mechanism. This complication is activated via pressing the pushpiece at 8:00. The wearer focus on the eye-catching complication at the top left corner of the dial. Two hammers, engraved with HW, are surrounded by a golden ring and this mechanism elegantly overlaps with the hours subdial, creating a harmonic scene. The low chimes of the repeater indicate hours as the high pitch indicates quarter hours and minutes. The minute repeater demands both technical and musical precision, and this model raises the bar for watches in its category.

JAEGER-LECOULTRE

MASTER GRANDE TRADITION MINUTE REPEATER – REF. 501.T4.50

With the interaction of its retrograde two-week power reserve and barrel-torque indicators on the intricately openworked dial, this audible member of the Master Grande Tradition collection is a comprehensive reflection of the house's vast range of horological expertise. Housed within a titanium case, the hand-wound 413-piece 947 caliber lets time resonate with perfect clarity and authority. Rich with handcrafted decorations, this timepiece oscillates at 21,600 vph and opts for an exceptional revealing design, with central dauphine hands, as the visual backdrop for its revered resonant complication. It is worn on an alligator leather strap.

JAEGER-LECOULTRE

REVERSO REPETITION MINUTES A RIDEAU – REF. 235.35.2M

Jaeger-LeCoultre introduces a limited series, dedicated to Venice, of the Reverso Répétition Minutes à Rideau, the first and only timepiece whose minute repeater function is activated by opening the curtain. A truly technical innovation, the metallic curtain is reminiscent of the original function of the reversible case: that of protecting the watch from impacts. This exceptional timepiece enables the Grande Maison to provide yet another demonstration of its expertise in the field of sound with the integration of trebuchet hammers, monobloc gongs made from a secret-formula alloy, and a silent regulator.

PARMIGIANI

TORIC CAPITOLE – REF. PFH276-1203300-HA1441

Immediately standing out with its ingenious sectorial display of the hours and minutes, this 18K white-gold timepiece, with polished finish and fully hand-engraved and -knurled bezel, serenades the exquisite sheen of its white mother-of-pearl dial with a two-gong "serpents" cathedral chime sounding the hours, quarter-hours, and minutes. Worn on an Hermès alligator leather strap, the watch is driven by the 397-component hand-wound PF 321 caliber with 45-hour power reserve. The demonstrative timepiece is finished with a sapphire crystal caseback engraved with the model's individual number.

PARMIGIANI

TORIC WESTMINSTER INDIGO – REF. PFH412-2002500

Piercing the hand-guilloché dial with "shaded beehive" motif, a remarkable tourbillon offers its revered companionship to a minute repeater that, upon activation via the sliding lever on the 46.6mm 950 platinum case, strikes the hours, quarter-hours, and minutes, with the symphonic distinction of the Westminster chimes. Topped with a sapphire cabochon that resonates with the visual appeal of the indigo lacquered face, this finely engraved timepiece boasts, in addition to the exposure of the sonorous hammers through a unique aperture at 9:00, a large date opening above the perfect center of the dial.

ZENITH

ACADEMY MINUTE REPEATER – REF. 18.2250.4043/01.C713

An iconic representative of the best that haute horology has to offer, this Academy Minute Repeater strikes the hours, quarter-hours and minutes once the dedicated repeater slide has been activated. A first hammer strikes the hours (1 to 12) on a low-pitched gong, while the quarters (1 to 3) are struck by two hammers on two low- and high-pitched gongs, and the minutes (1 to 14) by a hammer on a high-pitched gong. This highly complex El Primero 4043 movement comprises 461 parts housed within a rose-gold case that resonates to the sound of this subtle music of time.

Multi-Complications

A. Lange & Söhne
Lange 1 Tourbillon
Quantième Perpétuel

Wilhelm Schmid

CEO of A. Lange & Söhne

"The Internet is a playground of growing importance for luxury watch brands"

DUBAI BOUTIQUE

Wilhelm Schmid **LEADS ONE OF THE MOST CLASSIC AND TRADITIONAL MANUFACTURES** in the watch-making industry. In order to develop products and marketing that speak to Generation Y and younger people, Schmid is increasing A. Lange & Söhne's presence on social media networks.

How did this year finish for A. Lange & Söhne?

2012 was an excellent year for our business in all markets and a year full of highlights too. It started off very well with a fantastic fair in Geneva. Thanks to a diversified model range and brand message that is globally understood we have been facing a very pleasant market development. We opened five new boutiques and we decided to enlarge our manufacture with a new building.

How satisfied are you? With what are you not?

On a scale from 1 to 10 I would rate my satisfaction with a 9—to leave some room for improvement.

What do you expect from 2013?

We will not cease to measure ourselves against our personal best so that we will continue to expand and strengthen our position in the market. Our new timepieces will once again demonstrate our performance in the field of challenging complications, in full respect of our unique tradition.

What are the strengths and weaknesses of A. Lange & Söhne today?

A. Lange & Söhne's strength is based on the passion of our employees, our retail partners and the loyalty of watch collectors for the brand. This is our most valuable asset. Our biggest challenge, in an overall positive market environment, lies in our limited production capacity. Given the craftsmanship, effort and time used on each of our timepieces, we are simply not always able to meet the demand of Lange connoisseurs around the globe.

Who is the typical A. Lange & Söhne customer?

Our customers are independent-minded and quality-conscious individuals who share an affinity for fine mechanical timepieces and for whom only the best is good enough.

What is your attitude about social networks?

Social networks have a significant voice among luxury consumers. The future is digital and social media is a part of it. The Internet is a playground of growing importance for luxury watch brands as most of our end costumers have some kind of presence online.

You are present and active on Facebook, much less so on Twitter. What is the reason for that?

We must concentrate our efforts on the social networks that are most relevant for our brand. Therefore, we follow our end customers who are more active on Facebook.

Is luxury compatible with social networks?

Absolutely, on the one hand our customers may spend their time on Facebook or other networks, on the other hand they cherish the lining of a bespoke suit or the sophistication of a handcrafted mechanical watch. To the upcoming generation of our customers these are complementary aspects of their lifestyle.

Is it possible to sell A. Lange & Söhne watches over the Internet? What is your position on e-commerce?

The Internet has a huge significance for the formation of opinions. But it has its limits when it comes to personal experience and advice. Our customers are very happy with the distribution system we have, which relies on personal assistance on the spot.

What are the main markets of A. Lange & Söhne and how are sales allocated?

Our watches are available in 55 countries around the world with Europe, Asia Pacific and North America as the strongest regions.

What are your priorities today?

In our distribution, we have a dual focus: we have a very loyal customer base in our established markets, which we want to expand. In our new markets, we have growing fan communities, which need our attention. Internationalization is still one of our most important issues. The development of our brand in China, India, South America and in the Middle East region will continue to keep us busy in the years to come.

You have ten boutiques worldwide, including six in Asia and the Middle East. Are further store openings planned in these areas? And in Europe, North or Latin America?

Yes, we have ten boutiques around the world. However, our production capacity is limited in order to maintain the quality of our timepieces. Therefore it is not our aim to open multiple retail locations but to enhance consumers' experience of the brand in our key markets. For the same reason, we opened five new mono-brand stores in 2012—in Abu Dhabi, Dubai, Singapore, Palm Beach and Paris.

What is your view on watchmaking today?

Strictly speaking, mechanical watches are an anachronism. But one of the most successful and fascinating of our time. There is still so much to be discovered and developed—that is why fine watchmaking never gets boring.

What are the main challenges of this industry?

To keep the spirit of watchmaking alive in the future and develop new products and communications that speak to the younger target groups of Generation Y and beyond.

Gaëtan Guillosson

President of A. Lange & Söhne North America

"The Grande Complication will be this year's star"

CONSTRUCTING THE BEST WATCHES IN THE WORLD has always been the objective of the brand A. Lange & Söhne. This year, the German watchmaker launches the Grande Complication, a watch that contains no fewer than seven complications, making it the most complicated piece ever produced by A. Lange & Söhne.

If you had to explain A. Lange & Söhne to a neophyte in a few words, what would you say?
Rooted both in the Saxon watchmaking and the German engineering tradition, the name A. Lange & Söhne stands for unmatched craftsmanship and ingenuity that have often defined benchmarks for others. By tradition it is a unique brand that follows its own philosophy and in doing so constantly tries to outperform its own achievements.

How is A. Lange & Söhne different from other luxury watch brands? Its features?
A. Lange & Söhne has a long-standing history of setting milestones in the development of fine watchmakingand created a style of its own already in 1845. Today, we still adhere to the same principles unique to our tradition. The vision for A. Lange & Söhne is the same that has led the brand ever since it was founded: to build the world's best watches. Its cornerstones are the three-quarter plate made of untreated German silver, movement parts decorated or engraved by hand, use of the most precious materials like gold and platinum for the cases, and a timelessly elegant design. Above all, exciting and innovative technical features are of utmost importance for Lange. There are many examples for the Saxon ingenuity of our master watchmakers, like the outsize date, the tourbillon stop mechanism, the 31-day power reserve, the fusée-and-chain power transmission or the peripheral month ring.

What are the demands of U.S. customers?
U.S. customers have a heightened appreciation of German engineering and craftsmanship. Americans look for the highest degree of quality and performance in luxury purchases; it is what they find in our timepieces.

What is the position of A. Lange & Söhne on the North American market?

In our global distribution strategy, the American market has always played an important role. We have strengthened our presence by focusing on our retail partners: we have tightened our distribution to the best points of sale in the top cities. And we have opened our first A. Lange & Söhne Boutique in North America on Worth Avenue in Palm Beach.

GRANDE COMPLICATION

What is the most iconic model for today's brand?

The style of A. Lange & Söhne can be characterized as contemporary, classical elegance that stands the test of time. A. Lange & Söhne's design milestones include iconic timepieces like the Lange 1, with the first outsize date in a regularly produced wristwatch, and the Lange Zeitwerk, with its perfectly legible, precisely jumping numerals: both of them featured a new look in the history of watch design and were received with high esteem by our customers and the media.

Overall, how would you say that the brand has evolved over the past five years?

Due to the technical achievements of the timepieces, awareness of A. Lange & Söhne has heightened. In response to new movement developments there has been more international recognition, such as winning the L'Aiguille d'Or award. A. Lange & Söhne has also encountered increased record sales at auctions. At the main watch auctions, models such as the Tourbillon "Pour le Mérite," the 1815 Moon Phase or the Anniversary Langematik are fetching up to five times their original retail price. Therefore, leading auctioneers consider A. Lange & Söhne as one member of an exclusive group of watch brands that are expected to generate a long-term increase in value.

What are the strong points—products, events—for A. Lange & Söhne in 2013?

The hero of the year will definitely be the Grande Complication, the most complicated wristwatch ever produced by A. Lange & Söhne. With seven horological complications, including grand sonnerie, petite sonnerie, minute repeater and flying seconds, it is an exceptional timepiece at the pinnacle of fine watchmaking. It will be presented in more depth at different worldwide events later this year. We added two new members to our 1815 watch family, which pays tribute to the style elements of traditional Lange pocket watches: The new 1815 Up/Down with power reserve indicator and the 1815 Rattrapante Perpetual Calendar combining a split-seconds chronograph with a perpetual calendar. In the Grand Lange 1 "Lumen," a semi-transparent dial exposes the ingenious mechanism of the first luminescent outsize date.

1815 UP/DOWN

GRAND LANGE 1 "LUMEN"

Greg Simonian

CEO of Westime

"With our unique product selection, we put our customers on a path to discovery."

WITH 25 YEARS IN THE LUXURY WATCH BUSINESS, Westime is an undisputed leader in the field. Westime President Greg Simonian has grown up in the industry, getting a firsthand look at the sea changes that have occurred, with a unique perspective on the changes yet to come.

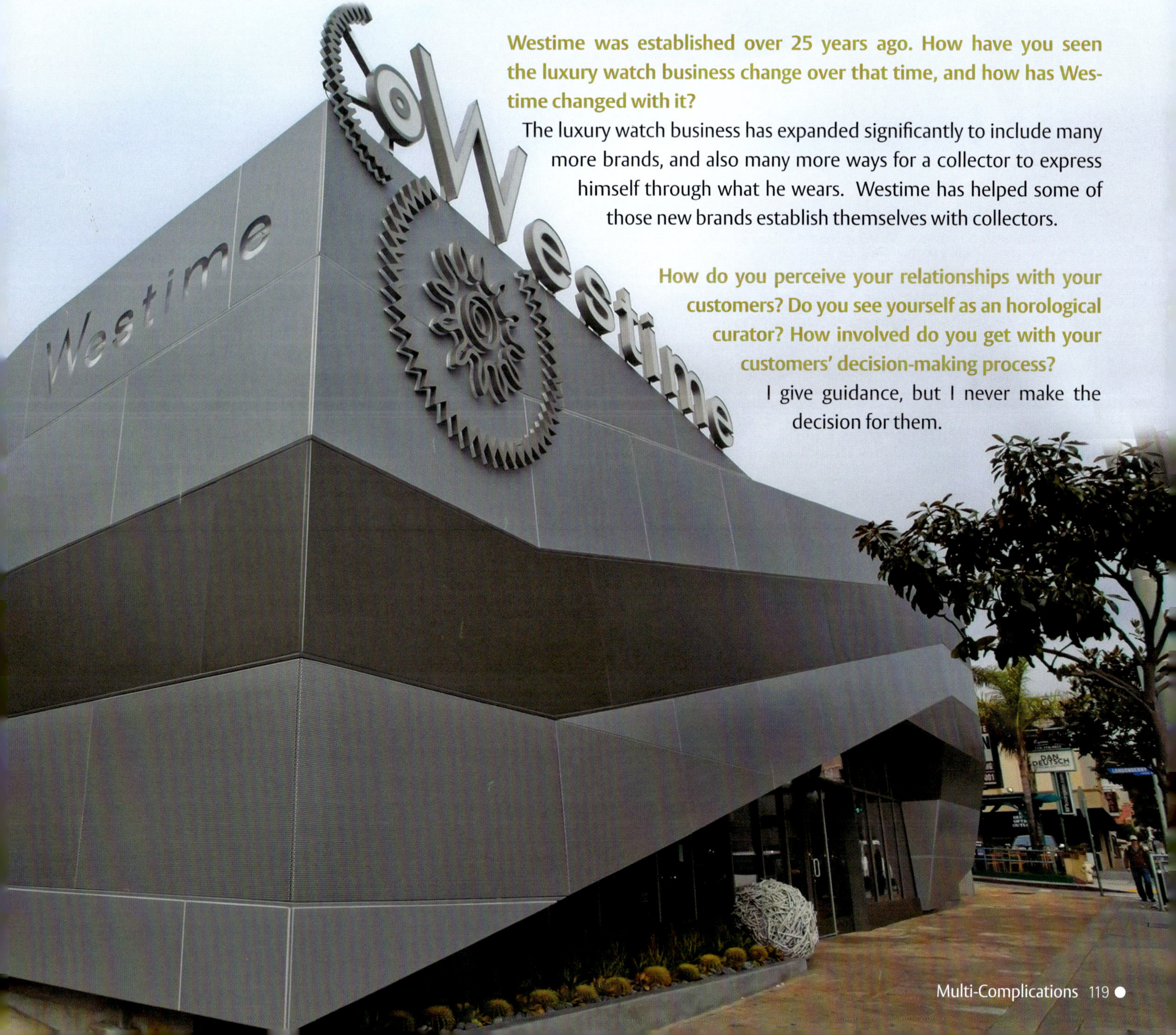

Westime was established over 25 years ago. How have you seen the luxury watch business change over that time, and how has Westime changed with it?

The luxury watch business has expanded significantly to include many more brands, and also many more ways for a collector to express himself through what he wears. Westime has helped some of those new brands establish themselves with collectors.

How do you perceive your relationships with your customers? Do you see yourself as an horological curator? How involved do you get with your customers' decision-making process?

I give guidance, but I never make the decision for them.

Would you say your customers are interested newcomers to the world of horology, or do people generally come in with a strong idea of the watch for which they are looking?

We get both. And we carry a wide selection of watches that are appealing to both types of customers. Additionally, our staff members know how to work with both.

How does Westime's status as a family business affect its role in the increasingly consolidated luxury watch world?

Family-owned stores are still the only places where you can go to see multiple brands in one location. Also, they are the only places where you can go to talk to the owner, and see a person standing behind a product rather than a corporation. I don't choose to carry watch brands because they may fit a certain price point, or another reason that a corporation may dictate. I choose each watch because I like it, and I think it would be interesting to our customers. Our stores provide a much more personal experience and I think people still appreciate that.

Westime has a location on Sunset Boulevard, in Beverly Hills, and in La Jolla. How would you describe the atmosphere of each boutique?

With our unique product selection, we put our customers on a path to discovery. Westime Beverly Hills is like a jewel box, with very special limited editions. Westime Sunset is much bigger, and more like an art gallery in terms of display. And Westime La Jolla is an extension of our Sunset store in a way, but in the San Diego area.

What is your approach toward selecting brands for Westime's retail locations?

I like to think of what's interesting to my customer, not a price segment, and to create a selection that puts our customers on a path to discovery at all price points. It is important for us to offer both watches for people who know what they want, and options for those who are willing to step outside the comfort zone.

What complications, materials or styles do your customers most often request?

Among complications, any innovative complication is becoming more popular, such as Urwerk's revolving satellites or Hublot's Key of Time. I am also seeing more interest in perpetual calendars. As for materials, whatever is new and innovative is definitely our customer. Although classic watches are part of every connoisseur's collection, big bold watches have always proven popular with Westime's clients.

What is your personal favorite innovation?

I love the Richard Mille RM 012 and Urwerk's rotating satellite complication. And for material innovation alone, it has to be the Richard Mille watches created for Rafael Nadal.

Single-brand boutiques are experiencing a rapid increase in popularity. How does this affect you, and what do you see as its effect on the future of the retail landscape?

I think it is positive for everybody. The increase in stores serves to educate more people about the world of watches. And there is still a need for multi-brand experts like Westime. Just consider, if you are going to buy a suit, do you go to a place that only sells one suit, or to a place that sells many? As the single-brand boutiques open, they open near one another and near us, giving customers an exciting shopping environment!

What is the most important part of your job?

Making sure the stores remain interesting for everyone. It is key that customers can turn to Westime for a watch in all price categories, and during all points in their lives.

Since the birth of mechanical watchmaking in the 14th century,

artisans have always striven to design instruments containing the greatest possible number of functions—first in larger clocks, then in smaller ones, before moving on to pocket watches and, more recently, to wristwatches. Today more than ever, these ultra-complicated pieces represent the quintessence of the horological art, the supreme demonstration of savoir-faire. Boosted by the progress accomplished in the field of dedicated computer software, which has fostered considerable advances in terms of miniaturization and reliability, the foremost horologers compete to develop "the most complicated watch in the world." Alongside these showpieces, created to set records, watchmakers offer collectors and lovers of rare pieces a wide choice of multi-complications associating three or four ultra-sophisticated functions, as well as marvelous astronomical watches.

Horology did not evolve in a purely linear fashion from simplest to most complicated timepieces. The minute display, which today seems one of the most basic functions, only became widespread at the end of the 17th century, with the invention of the regulating balance spring. However, some pocket watches already featured highly complex mechanisms by this point. As for the seconds display, it would not come along until another century later. Moreover, the history of horology is marked by several "setbacks," due to the shrinking size of the mechanisms. This is why the first portable clocks and pocket watches of the Renaissance were more rudimentary, technically, than the large astronomical clocks of the Middle Ages—and even the most sophisticated wristwatch today would not be able to display as many functions as an ultra-complicated pocket watch.

The first explicit mentions of mechanical clocks appear at the beginning of the 14th century. Though we can reasonably assume that artisans began with simple mechanisms, the only clocks we know from that period are quite complex, such as the clock tower with the astronomical dial that Roger Stoke constructed for the Norwich cathedral between 1321 and 1325, and of course the "star" of the era, a marvel of its time, Giovanni de Dondi's astrarium in Padua, completed in 1364.

During the late Middle Ages and the Renaissance, the Church and political authorities encouraged and financed monumental clocks that boasted a large number of astronomical and astrological indications, equipped with audible strike systems that were often quite spectacular. Among the most famous examples are the astronomical clocks of the Lund cathedral, in Sweden (1380), the Saint-Etienne cathedral in Bourges, France (1424) and the Saint Mary of Rostock church in Germany (1472).

The first pocket watches, at the end of the 15th century, possessed only an hour hand. In the 16th century, artisans developed astronomical and astrological displays, then added more and more functions—alarm, calendar, strike in passing and on demand—as they competed to create watches that contained the largest number of complications.

The most famous example of this phenomenon is the Breguet N° 160, better known as the "Marie-Antoinette," whose creation and history are shrouded in history. In 1783, Abraham-Louis Breguet, already a favorite of Louis XVI and his wife, received a commission from an officer of the queen's guards: to create a watch that would integrate all the complications and refinements of the time. Marie Antoinette, guillotined in 1793, would never see this beauty, as it was completed in 1827, under the direction of Breguet's son. This automatic-winding watch (referred

to as a perpétuelle) contained within its gold case and rock crystal dial complications such as a minute repeater, complete perpetual calendar, equation of time, power reserve indicator, metallic thermometer, independent large seconds and small seconds hand—not to mention the many technical feats such as a double pare-chute, a gold balance spring and several components in sapphire crystal. The Marie-Antoinette passed through several hands, and was stolen from a Jerusalem museum in 1983. The company Breguet created an identical reproduction of the timepiece, but in the autumn of 2007, the original suddenly reappeared in the museum's collections under mysterious circumstances.

In 1925, the New York banker Henry Graves, Jr. commissioned Patek Philippe to create for him the most complicated watch ever. Comprised of over 900 components, this double-sided pocket watch includes 24 complications, including a grand strike on four gongs and alarm on a fifth gong. The title of "the world's most complicated watch" was long disputed between the Graves and another horological masterpiece, the Leroy 01 (1904). Its double-sided case contains 20 complications related to time measurement—equation of time, chronograph, the sky above the Northern and Southern Hemispheres, grand strike, etc.—and five non-horological functions: hygrometer, thermometer, barometer, altimeter and compass. This inclusion has given rise to a dispute between those who grant it 25 complications and those—partisans for the Graves—who contest that it has "only" 20.

In 1989, Patek Philippe put an end to the quarrel with its Calibre 89. With 33 complications and 1,728 components, this timepiece is today the most complicated portable watch in the world, though describing it as "portable" might be a stretch—at 88.2mm across and 41.07mm thick, weighing over 2 pounds, the Calibre 89 is not exactly a pocket or fob watch.

The advent of the wristwatch in the early 20th century obliged watchmakers to restrict their ambitions a bit for complicated pieces, given the smaller volume of the wristwatch's case. However, it also inspired them to come up with treasures of ingenuity to miniaturize existing complications and combine them in exceptional pieces. Today, the biggest names in the industry compete to create "the most complicated watch in the world," aided in their quest by computer software that allows for possibilities still unimaginable a few decades ago. In March of 2012, Marc-Alexandre Hayek unveiled the prototype of a Breguet wristwatch offering 35 different functions on the piece's two faces. Named Hommage à Nicolas G. Hayek Ref. 7887, this piece has what it takes to become "the most complicated wristwatch in the world." Stay tuned.

Pocket Records

Though pocket watches have been largely replaced by wristwatches, their larger cases provides horologers with a versatile field of experimentation when it comes to gathering as many complications as possible within the same timepiece. However, without shooting for the Guinness Book of World Records, larger horologers as well as artisans have recently offered extremely sophisticated pocket watches.

Cartier
Grande Complication Squelette Pocket Watch
An ode to transparency, the Grand Complication Squelette Pocket Watch from Cartier associates a tourbillon, a monopushbutton chronograph, a perpetual calendar and an eight-day power reserve within its finely openworked 457-component movement.

Vincent Bérard, *Quatre Saisons Carrosse*
Artisan watchmaker Vincent Bérard has chosen to reinvent the carriage watch with the Quatre Saisons Carrosse. This handmade piece combines a perpetual calendar, quarter repeater, a system of automatons on the dial, a power reserve indicator and a thermometer.

Breguet, *Nº 5*
The Nº 5 from Breguet, a replica of an older watch, associates automatic winding, 60-hour power reserve, moonphase display and a quarter repeater whose hammer strikes the inside of the case instead of a gong.

Patek Philippe, *Calibre 89*

With 33 complications and a 1,728-piece movement, the Calibre 89 from Patek Philippe is to this day the most complicated portable watch in the history of horology. Besides the complete roster of "classic" complications (tourbillon, calendar, chronograph, strike, astronomical complications, etc.), it presents a few extremely rare functions such as the "secular" perpetual calendar (see chapter on Calendar watches), as well as a display of the date upon which Easter will next fall. It is also equipped with a celestial map that depicts the Milky Way and 2,800 stars of the Northern Hemisphere, as well as their relative size. Its development and production required nine years.

Patek Philippe, *Star Caliber 2000*

The double-sided pocket watch Star Caliber 2000, from Patek Philippe, associates 21 complications. Dedicated to our sun, the obverse dial displays functions including a running equation of time, times of sunrise and sunset, and a 24-hour day/night indication. The second dial, on the reverse side, pays tribute to the night sky with a map featuring the moon and stars, a moonphase display and a golden ellipse outlining the part of the sky visible from the selected observation site.

Breguet, *Modern Reproduction of the Marie-Antoinette*

In 2008, Breguet presented an identical reproduction of the famous ultra-complicated Marie-Antoinette watch (1827), one of the most impressive achievements of its history, housed in a box carved from the queen's favorite oak at Versailles.

Audemars Piguet, *Grande Complication Savonnette*

Within its yellow-gold case cover, the Grande Complication Savonnette (hunter-type watch) from Audemars Piguet unites a perpetual calendar, minute repeater and split-seconds chronograph.

Exploits on the Wrist

The race to produce "the most complicated watch in the world" is in full swing. To stand out from the crowd, the most high-end watchmakers come up with feats of ingenuity, reinventing the base mechanisms that support the complications, most of which were invented and perfected before the 19th century. The time has come for new constructions, in new materials, with new interpretations that playfully overturn long-established traditions.

Patek Philippe, *Sky Moon Tourbillon*

The most complicated Patek Philippe wristwatch, the double-sided Sky Moon Tourbillon (2001) combines 12 complications, including a minute repeater with cathedral gongs, a perpetual calendar with retrograde date hand and a tourbillon. The dial on the reverse side presents a mobile map of the sky, reproducing—with extreme precision—the visible movement of the stars and the moon, as well as the moonphases.

Dominique, *Loiseau 1F4*

Produced in just two models per year, the 1F4 from Dominique Loiseau associates numerous functions, including a perpetual calendar, split-seconds chronograph, second time zone, equation of time, flying tourbillon, minute repeater and grand and small strike.

Franck Muller, *Aeternitas Mega 4*

The Aeternitas Mega 4 model from Franck Muller houses, within its elegant tonneau-shaped case, 15 complications, including grand and small strike, minute repeater with Westminster chime, secular calendar with retrograde date, equation of time, large tourbillon, split-seconds chronograph and two additional time zones.

Vacheron Constantin, *Tour de l'Ile*

In 2005, Vacheron Constantin celebrated its 250th anniversary by creating the ultra-complicated model called Tour de l'Ile. This double-sided timepiece contains in its 834-piece movement no fewer than 16 complications, including a minute repeater, tourbillon, indications of the moonphase and age of the moon, perpetual calendar, equation of time, times of sunrise and sunset, and a map of the night sky.

Jaeger-LeCoultre

Hybris Mechanica

The 2009 Hybris Mechanica trilogy from Jaeger-LeCoultre displays a total of 55 complications, some of which are found on more than one model. The Hybris Mechanica a Gyrotourbillon is an homage to the quintessence of the mechanical measure of time, thanks to its "spherical tourbillon" (see chapter on Tourbillons), completed by a perpetual calendar with four retrograde hands and an equation of time function that can be adjusted according to one's location. The Hybris Mechanica à Tryptique comes up with an unusual take on the famous Reverso reversible case, originally created in 1931. The watch boasts three faces, for three dimensions of time. On the front of the case is civil time, with a new detent escapement system and a titanium tourbillon; the back shows sidereal time, with a sky map, Zodiac calendar, equation of time and times of sunrise and sunset; while the watch cradle displays perpetual time, with a complete calendar—making a grand total of 18 complications on this exceptional timepiece. As for the Hybris Mechanica à Grande Sonnerie, it contains no fewer than 26 complications, with a 1,300-piece movement that powers a flying tourbillon, perpetual calendar with retrograde hands and a strike/minute repeater mechanism with numerous patented innovations.

Hybris Mechanica à Grande Sonnerie

Hybris Mechanica à Gyrotourbillon

Reverso à Tryptique Hybris Mechanica

Varied Triumphs

The horological superstars whose complications can be counted on the fingers of two, three, four, even five or six hands should not make us forget all the creations that—though they have achieved more modest recognition—are exceptional triumphs in their own right. A "grand complication," in the watch world, is created when a watch contains three or four supplementary functions—especially if each of those functions belongs to one of the traditional domains of haute horology (audible indications, calendars, chronographs, tourbillons, etc.).

Zenith,
Grande Class Traveller Répétition Minutes El Primero

Composed of 744 pieces, the movement that powers the Grande Class Traveller Répétition Minutes El Primero, from Zenith's 2005 releases, combines a minute repeater, alarm (with audible or vibration settings), a second time zone, a large date with instant jump and two power reserves, one for the watch, the other for the strike—not to mention, of course, the famous El Primero chronograph functions.

Jean Dunand
Grande Complication

In 2005, the brand Jean Dunand—co-founded by Christophe Claret, the renowned constructor of complicated movements—presented a wristwatch called Grande Complication. This 827-piece timepiece includes, among other functions, a single-pusher split-seconds chronograph, tourbillon, minute/quarter/hour repeater and perpetual calendar with retrograde date and day indications.

Vacheron Constantin
Patrimony Traditionnelle Calibre 2755

A worthy descendant of the prestigious Tour de l'Ile, released for the 250th anniversary of the brand, the Patrimony Traditionnelle Calibre 2755 from Vacheron Constantin brings together three major complications of the horological universe—minute repeater, tourbillon and perpetual calendar—in an elegant round case.

Patek Philippe
Triple Complication Ref. 5208

The automatic Triple Complication Ref. 5208 wristwatch from Patek Philippe associates within its platinum case a minute repeater, single-pusher chronograph and instantaneous perpetual calendar with apertures. It also stands out for its escapement and balance spring in Silinvar®, a silicon derivative.

Cartier, *Rotonde Grand Complication Skeleton*

Issued in a limited edition of 30 pieces, Cartier's Rotonde Grand Complication Skeleton boasts a tourbillon, perpetual calendar, single-pusher chronograph and eight-day power reserve display—all within an entirely openworked, decorated 457-piece movement.

IWC, *Portuguese Grand Complication*

The Portuguese Grand Complication from IWC offers a minute repeater and chronograph, as well as a perpetual calendar that indicates not only the date, day and month, but the year as well, via a four-digit display.

HD3, *Slyde*

An ultra-modern, ultra-geeky version of grand complications, the new electronic model Slyde from HD3 is distinguished by its curved tactile screen that displays the various indications (time, calendar, chronograph, countdown clock, moonphase, second time zone) in much the same way as a smartphone.

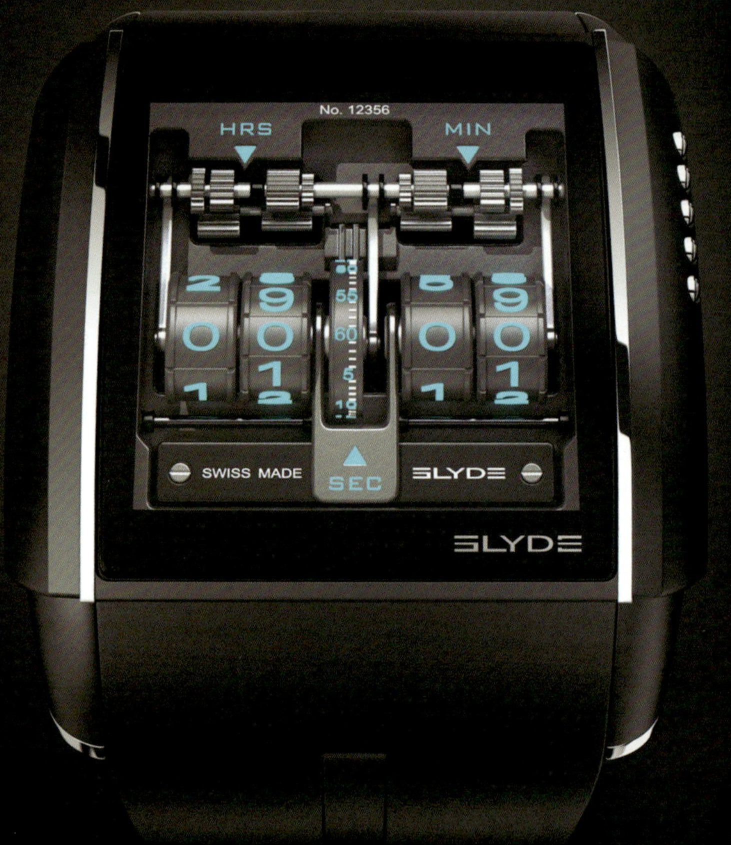

Christophe Claret, *DualTow NightEagle*

Christophe Claret scored a huge coup with its DualTow NightEagle, which features hour/minute displays on two belt drives inspired by treaded vehicles, a single-pusher planetary chronograph with strike and tourbillon, not to mention many other refinements and technical innovations.

Maîtres du Temps, *Chapter One*

Presented in 2008 in a tonneau-shaped case and rereleased in 2010 in a round version, Chapter One from Maîtres du Temps unites a single-pusher column-wheel chronograph, a retrograde second time zone display and retrograde date, as well as day and moonphase displays on two rollers at six o'clock and twelve o'clock.

Jean Dunand, *Shabaka*

Boasting a highly original, Art Deco-inspired dial, Shabaka from Jean Dunand associates a minute repeater with cathedral chime with a complete perpetual calendar with instantaneous day/date/month displays on rollers.

Patek Philippe, *Ref. 5216*

Reinterpreting one of the manufacture's classics, Patek Philippe's Ref. 5216 conceals beneath its sober dial a minute repeater, tourbillon, perpetual calendar with retrograde date hand, and a moonphase display.

Audemars Piguet

Royal Oak Offshore Grande Complication

Equipped with highly visible architecture that reveals its openworked movement, the Royal Oak Offshore Grande Complication brings together a perpetual calendar, minute repeater and split-seconds chronograph within its titanium and ceramic case.

A. Lange & Söhne

Lange 1 Tourbillon Quantième Perpétuel

The Lange 1 Tourbillon Quantième Perpétuel from A. Lange & Söhne associates its two classic complications in a design that is unmistakably recognizable as the Lange 1, with its off-centered hours/minutes display and a large date in a double aperture.

Astronomical Watches

Astronomical complications are the very source of the art of horology, and they are part of the classic repertory of all ultra-complicated pieces. At the start of the third millennium, they are making a forceful comeback on highly sophisticated pieces that meld technical prowess and poetry. The major rising star is the map of the night sky.

Van Cleef & Arpels, *Midnight in Paris*

The Midnight in Paris model by Van Cleef & Arpels reproduces on its aventurine dial the movement of the stars in the sky above Paris, the City of Light. The disc around the calendar on the caseback is set with meteorite.

Louis Moinet, *Astralis*

Besides its astral tourbillon and split-seconds column-wheel chronograph, the Astralis from Louis Moinet—a limited edition of just 12 pieces—stands out for its 24-hour planetary function, which successively displays four planets, each composed of an extremely rare meteorite.

Ulysse Nardin, *Trilogy of Time*

Ulysse Nardin has demonstrated its commitment to the astronomical and astrological science of the Renaissance by creating an exceptional trio of timepieces called Trilogy of Time. The first piece, Astrolabium Galileo Galilei, depicts on its dial the functions of the astrolabium, a navigational tool that facilitated great discoveries. It indicates the respective positions of the sun, moon and stars as observed from the Earth. The Planetarium Copernicus continuously displays the month, Zodiac sign and astronomical positions of the planets in relation to the sun and the Earth using a dial composed of seven concentric rings. As for the Tellurium Johannes Kepler, with its cloisonné enamel dial representing the Earth as seen from above the North Pole, it displays times of sunrise and sunset (with areas of shadow and light), as well as solar and lunar eclipses.

Audemars Piguet
Royal Oak Equation of Time

Equipped with an equation of time system that takes longitude into account (see chapter on Equations of Time) the Royal Oak Equation of Time from Audemars Piguet is also one of the rare wristwatches to display times of sunrise and sunset utilizing three parameters: date, longitude and latitude.

Patek Philippe, *Celestial Ref. 5102*

The Celestial Ref. 5102 from Patek Philippe continuously displays the exact configuration of the night sky in the Northern Hemisphere, with the apparent movement of the stars, the position of the moon and its phases over the course of the lunar cycle.

IWC, *Portuguese Sidérale Scafusia*

The most complex watch ever created by IWC, the Portuguese Sidérale Scafusia displays on its double-sided case the solar time, sidereal time (that of the stars) and various other astronomical indications including a map of the night sky and times of sunrise and sunset—not to mention a constant-force tourbillon and perpetual calendar.

Jaeger-LeCoultre
Master Grande Tradition Grande Complication

On the Master Grande Tradition Grande Complication from Jaeger-LeCoultre, the "orbital" flying tourbillon that serves as an hour hand indicates sidereal time, or time calculated by the stars. Civil time is indicated by a small sun turning around the periphery of the dial, and a Zodiac calendar indicates the position of the constellations as they appear throughout the year.

Chopard, *L.U.C All-in-One*

To celebrate its 150th anniversary, Chopard presented the L.U.C All-in-One model. Possessing 516 components, the L.U.C 4TQE movement—certified as a chronometer and marked with the Poinçon de Genéve—displays 14 complications (tourbillon, perpetual calendar, etc.) on the front and back of the watch, including, on the caseback, equation of time, power reserve, 24-hour day/night indication, times of sunrise and sunset and astronomical moonphase.

A. LANGE & SÖHNE

SAXONIA ANNUAL CALENDAR – REF. 330.025

The new platinum version of the Saxonia Annual Calendar lends a sophisticated shine to a model that became an instant classic upon its 2010 release. The piece's calendar indication includes the day, date, month and moonphase. Taking into account the variance between months of 30 and 31 days, the calendar mechanism need only be corrected at the end of February. Its prominent moonphase indication loses only one day every 122 years. A peek through the sapphire crystal caseback reveals the automatic-winding manufacture caliber L085.1 SAX-O-MAT, constructed to show off the classic screw balance—the watch's beating heart.

AUDEMARS PIGUET

JULES AUDEMARS OPEN-WORKED TOURBILLON CHRONOGRAPH REF. 26353PT.OO.D028CR.01

The Jules Audemars Open-worked Tourbillon Chronograph is powered by Audemars Piguet's in-house Manufacture caliber 2889, which is comprised of 290 parts and beats at 21,600 vph. This timepiece has a platinum case featuring a fixed bezel, lugs, a faceted crown and two pushbuttons at 2:00 and 4:00. It fascinates by the sheer complexity of its open-worked dial bearing a tourbillon carriage at 6 o'clock along with subdials. Its two blued steel hands and a slender sweep seconds hand glide around the blue Arabic numerals appearing on the inner bezel ring. The strap is large-scale alligator leather.

AUDEMARS PIGUET

ROYAL OAK OFFSHORE GRANDE COMPLICATION – REF. 26065IS.OO.1105IS.01

The Royal Oak Grande Complication is powered by the 2885 caliber, a self-winding movement of 13.5''', which features a split-seconds chronograph, a minute repeater and a perpetual calendar. Seconds are displayed in a subdial at 9:00. Housed in an 18K white-gold case, the movement beats at 19,800 vph. The dial displays 12 indications, including day, date, month, leap years and moonphase. The movement is comprised of 654 pieces.

AUDEMARS PIGUET

TRADITION MINUTE REPEATER TOURBILLON CHRONOGRAPH REF. 26564IC.OO.D002CR.01

The Tradition Minute Repeater Tourbillon Chronograph houses the hand-wound, in-house Manufacture caliber 2874, which is enclosed in a titanium case with a white-gold bezel, crown, pushbuttons and caseback. The dial is silvered opaline and has pink-gold applied Arabic numerals and hands. The movement is comprised of 504 parts and beats at 21,600 vph. Its functions include a minute repeater, tourbillon, chronograph, hours, minutes and small seconds. The minimum power reserve is 48 hours. The timepiece is not water resistant. This is a limited edition of 10 pieces.

CHRISTOPHE CLARET

21 BLACKJACK – REF. MTR.BLJ08.370-391

A real miniature casino, the 21 Blackjack matches grand complications with the world of gaming. To start there is Dice, which features a pair of miniature dice located in a cage at 4:00. Secondly there is a game of roulette on the back of the watch and the winding rotor also functions as the roulette wheel. Lastly a game of Blackjack is on the lower part of the dial, between 3:00 and 9:00 where the player's four cards appear in windows. On the upper part of the dial are three additional windows for the dealer's cards. Each time a shutter opens, whether for the player or the dealer, there is a gong, whose mechanism is visible through a side window at 2 o'clock.

CHRISTOPHE CLARET

BACCARA – REF. MTR.BCR09.120-129

Christophe Claret presents the Baccara, a spectacular timepiece that doubles as a miniature casino with three games: baccarat ('baccara' in French), roulette and dice. The mechanical wonder that is the exclusive movement—which required over a year of development—is complemented by the aesthetic wonder of the special sapphire crystal protecting Baccara's dial, which when blown upon reveals a secret, talismanic Chinese ideogram. On the back of the watch, the winding rotor, visible through a sapphire crystal, serves as a roulette wheel. Once set in motion by one or two movements, the wheel turns for a few moments before stopping. The Baccara forms part of the original field of horological innovation initiated by Christophe Claret: that of playful, interactive timepieces.

DIOR

CHIFFRE ROUGE C01 – REF. CD084C10A002

Available in a limited edition of 200 pieces, Dior watch Chiffre Rouge C01 model is both technical and elegant. Equipped with several small complications—displays for power reserve, day of the week and retrograde date—on a black dial whose "cotton piqué" pattern recalls the dickey on a formal shirt, this 38mm version is fitted with a hand-sewn black alligator strap matching the dial.

FRANCK MULLER
AETERNITAS MEGA

Simply put: the most complicated wristwatch ever made! With 25 of this masterpiece's 36 complications visible to the wearer's amazement, the Aeternitas Mega may make its most impressive claim in the clarity with which its multitude of complexities are displayed. With an array of retrograde counters, a highly intricate single-pusher split-seconds chronograph, an extraordinary eternal perpetual calendar, the choice of grand or small strike of the Westminster Cathedral carillon, and two additional 24-hour time zones, to name a few of the 36 modules, this 1483-component magnum opus demonstrates the infinity of Franck Muller's daring vision of haute horology.

GIRARD-PERREGAUX

1966 MINUTE REPEATER, ANNUAL CALENDAR AND EQUATION OF TIME REF. 99651.52.131.BKBA

Within its 42mm diameter, the Girard-Perregaux 1966 Minute Repeater, Annual Calendar and Equation of Time houses three of the most illustrious watchmaking complications. They are displayed on the dial in perfect harmony, orchestrated by a manual-winding mechanical caliber with over 400 components. Surrounded by solid gold hour markers upon a silvered sunburst dial, the various indications perform their functions with refined panache. The imposing date at 1:30 immediately catches the eye, with the month in a window opposite, completing the annual calendar function. The equation of time displays the difference between civil time and solar time at 4:30. To complete this watchmaking masterpiece, a minute repeater enables the user to hear the striking of the hours, quarter hours and minutes. The case is specially constructed and designed for optimal acoustics.

HARRY WINSTON

HARRY WINSTON MIDNIGHT TOURBILLON CHRONOGRAPH – REF. 450/MMTC42RL.W

Harry Winston's Midnight collection emphasizes technological innovation as well as visionary design. The classic tourbillon is paired with a contemporary chronograph, combining to create an unforgettable model. The openworked dial allows the wearer to view the tourbillon as it moves on the bottom left of the dial. Surrounded by a rose-gold ring, this complication overlaps the sleek, avant-garde chronograph. The hour and minutes are off-centered while the oversized numerals at 12:00 and 6:00 add an extra dimension to the dial. The chronograph minutes counter and seconds hand are deep blue, a signature shade at Harry Winston. Black, white, dark gray and sky blue finishes bring out the ultra-modern qualities of this classic piece.

HAUTLENCE

HL2.2

This collection is Hautlence's first to offer an automatic-winding manufacture movement. Developing this line required three patented achievements. The first patent concerns the half-trailing hours chain: every 60 minutes the chain advances in 3-4 seconds without jumps, revealing the beauty of the movement in action. The second relates to the regulating organ integrated into the mobile bridge movement: its constantly changing position compensates statistically for the effect of gravity, enhancing its precision. The third patent covers the movement's two barrels: the principal barrel, wound by the automatic system, transmits the energy for a second barrel devoted exclusively to the complication. The complex systems are thus powered independently without affecting the precision of the watch. This 18K 4N rose-gold model is available in a limited, numbered edition of 28 pieces, as is the HL2.1 in 18K white gold.

HUBLOT

CATHEDRAL MINUTE REPEATER TOURBILLON COLUMN WHEEL CHRONOGRAPH REF. 704.QX.1137.GR

A world first, this multi-complication masterpiece boasts a minute repeater, tourbillon and column-wheel chronograph, all housed within a carbon fiber case. As spectacular as the watch's range of competency, the manual-winding timepiece demonstrates all three complications on the exhibition dial. Featuring carbon fiber bridges, shock absorption technology and a flat hairspring, the HUB8100 caliber powers a flying tourbillon without ball bearing, two cathedral-type gongs for the minute repeater, and a single-pushbutton chronograph with 60-second and 30-minute counters. A magnificent blend of expert horological tradition and refined modernity, this 48mm triple-complication wristwatch has a power reserve of 120 hours.

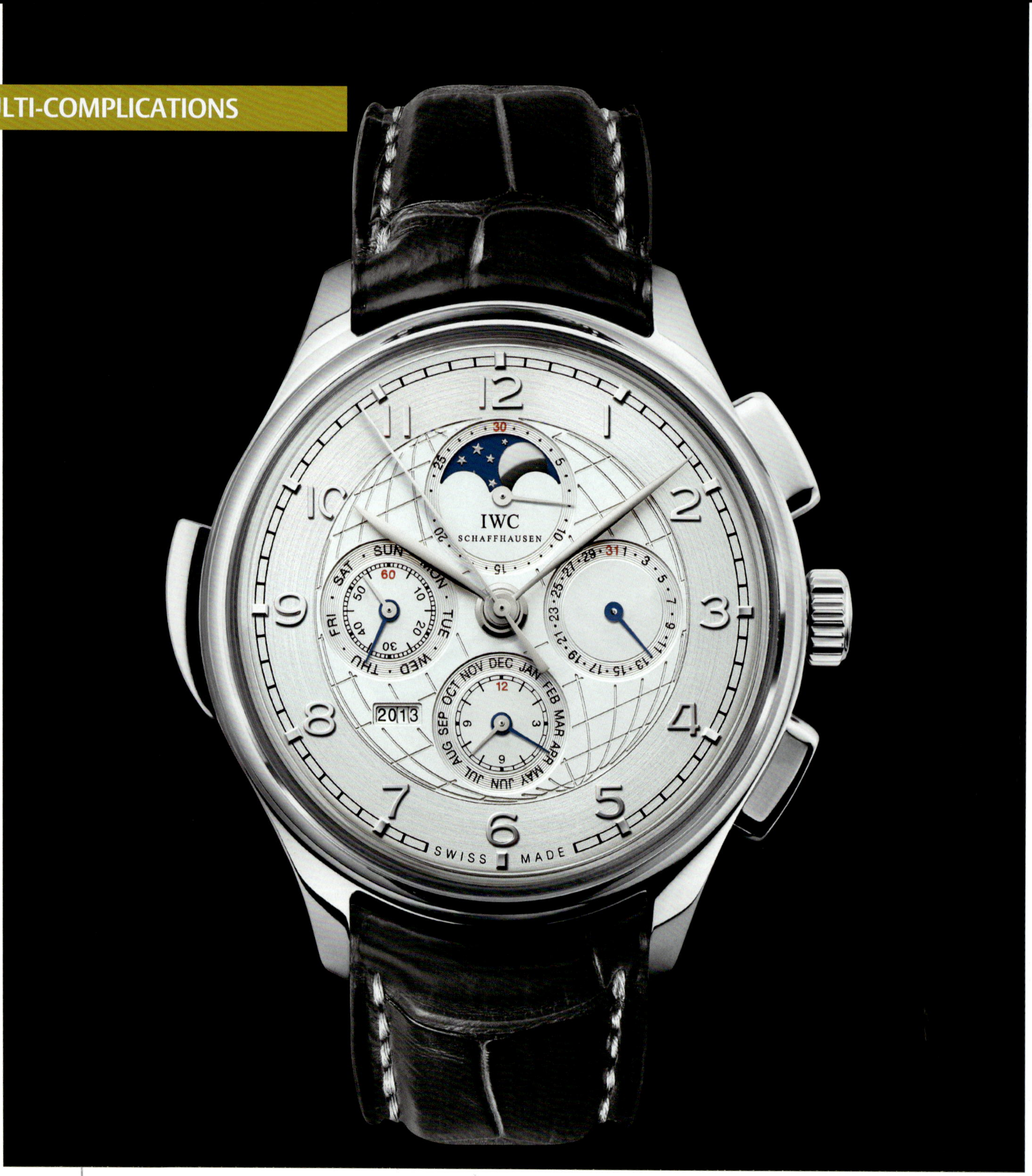

IWC

PORTUGUESE GRAND COMPLICATION – REF. IW377401

The Portuguese Grand Complication combines a multiplicity of extraordinary horological feats within a platinum case. The cool color of the case and black alligator leather strap, stitched with platinum threads, draws the eye to a cartographic white dial design that pays tribute to Vasco da Gama's navigational exploits. Along with a full perpetual calendar, programmed until the year 2499, and a small seconds subdial, this flagship masterpiece features a chronograph with central seconds hand, as well as 12-hour and 30-minute counters, a perpetual moonphase display and a minute repeater. On the outside of the case, the crown, chronograph pushers and repeater's sliding activator echo the brilliance of the dial's numerous appliqués. The theme of navigational exploration extends to the caseback, upon which a sextant is engraved.

IWC

BIG PILOT'S WATCH PERPETUAL CALENDAR TOP GUN – REF. IW502902

The Big Pilot's Watch Perpetual Calendar Top Gun is one of the most intricate pilot watches of all time. With a wide array of grand complications housed within, this sharp model's black ceramic Ø 48mm case and matching black dial present a bold exterior for an extremely functional model. The in-house manufactured movement complete with automatic-winding Pellaton 51614 caliber includes a perpetual calendar with a four-digit year display along with indications for the month, day and date. The double moonphase for the Northern and Southern Hemispheres guarantees that no pilot will get lost in the sky. Despite the complexities of this timepiece, the watch wearer can easily adjust its several functions via the screw-down crown at 3:00.

JACOB & CO.

CALIGULA LIMITED EDITION – REF. CAL1-RG

Jacob & Co. explores the playful and sophisticated tradition of erotic watches with the Caligula Limited Edition. This seemingly sober timepiece in 18K 5N rose gold conceals a secret: an erotic scene hidden beneath the dial. Upon turning the supplementary crown, part of the dial turns to unveil the scene, which is animated by rotating the crown positioned at 3:00. The scene and automaton were hand-painted by the miniaturist artist Andre Martinez: a true demonstration of craftsmanship requiring 15 hours of labor. The Caligula timepiece features a delicately hand-finished, automatic-winding Concepto Manufactured Calibre C-03025. A sapphire crystal caseback reveals the movement's exquisite finishing techniques: circular graining and Côtes de Genève, as well as sand-blasting and colimaçon finish for the rotor. This timepiece is available in a limited edition of 69 pieces.

JAEGER-LECOULTRE

HYBRIS MECHANICA A GRANDE SONNERIE – REF. 603.34.20

The Grande Sonnerie mechanism far surpasses the striking mechanism of any other complication. Jaeger-LeCoultre's exceptional mastery has created a magnificent movement that features a Grande Sonnerie with Westminster chimes, designed to ring out the longest melody ever played by any watch with a striking mechanism (the "Trébuchet Hammers," patent pending). This is all combined with a perpetual calendar and a flying tourbillon. Containing over 1,300 parts and the product of a five-year development process, the Jaeger-LeCoultre Calibre 182 is the most complex and most sophisticated movement ever created for a wristwatch to date.

JAEGER-LECOULTRE

MASTER GRANDE TRADITION TOURBILLON A QUANTIEME PERPETUEL – REF. 500.24.2A

Jaeger-LeCoultre's first tourbillon with a perpetual calendar has it all: an automatic proprietary caliber with two complications; a combination of high technical performance and the grandest watchmaking tradition; highly sophisticated finishes on the case; exceptional finishes on an entirely hand-decorated caliber. The movement meets demanding criteria of reliability, sturdiness and accuracy because of its outstanding elements: large balance wheel with variable inertia thanks to its adjustment screws, ultra-light titanium tourbillon carriage and high frequency of 28,800 vph.

PARMIGIANI

TORIGRAPHE – REF. PFH438-1202600

Upon a dial with an outer edge in sand-colored opaline, and center with Côtes de Genève decoration, four alluring apertures reveal the sophisticated inner workings of the hand-wound PF 356 caliber. Complementing its central seconds hand, the chronograph's snailed 30-minute counter, at 3:00, echoes, on a smaller scale, the visual character of the date subdial to its left at 9:00. At 6:00, the superb tourbillon is exposed, exemplifying the high-precision personality of this 45mm 18K white-gold timepiece with sapphire crystal caseback, luminescent hands and subtly integrated power reserve at 12:00.

PARMIGIANI

TONDAGRAPHE – REF. PFH236-1000300

Rich in aesthetic impact and mechanical sophistication, the Tondagraphe uses the intricacy of an open-worked dial to expose the prowess of its driving caliber, the self-winding PF 354. Engraved on the caseback, the 30-piece "Edition Numérotée" boasts a tourbillon at 12:00, quarter-second chronograph with 30-minute counter at 3:00, power reserve at 12:00 and rhodium-plated small seconds subdial at 9:00. Enclosed within a 43mm three-part case in 18K rose gold, the multi-complication wristwatch boasts a peripheral tachometer scale for the precise cross-referencing of time and distance against the speed of a moving object.

PATEK PHILIPPE

MEN'S GRAND COMPLICATIONS – REF. 5213G

This white-gold perpetual calendar's silvery opaline dial is a story of intelligent interjection and harmonious use of space. Flanking the retrograde date arc, two apertures indicate the day and month, while a third opening, discreetly positioned at 12:00 beneath the 16th day, lets the wearer know the status of the leap-year cycle. Also driven by the self-winding R 27 PS QR caliber, a two-gong minute repeater chimes upon activation via the sliding mechanism on the left side of the 40.6mm white-gold case. Combining the moonphase and small seconds, a subdial at 6:00 completes the distinctive design of this truly comprehensive multi-complication timepiece.

PATEK PHILIPPE

MEN'S PLATINUM GRAND COMPLICATIONS – REF. 5204P

Within its distinguished 40mm platinum case, the Men's Platinum Grand Complications makes a spectacular performance of time's perpetuity. Armed with a split-seconds chronograph for the measure of separate coinciding events, the hand-wound CHR 29-535 PS Q movement measures time with an oscillation frequency of 28,800 vph. On the right of the perpetual calendar's date display with moonphase interior, an aperture reveals the leap-year status. On the left, an opening indicates the day or night via a colored circle. Continuing the multiple layers of symmetry, the day and month are read side by side at 12:00 while two counters—small seconds at 9:00 and instantaneous 30-minute counter at 3:00—complete the thorough and sophisticated design.

PIAGET

PIAGET GOUVERNEUR TOURBILLON – REF. G0A37115

A member of Piaget's black tie collection, the Gouverneur Tourbillon combines elegant aesthetics with technical expertise. Entirely manufactured in house, this model is equipped with the ultra-thin, hand-wound 642P caliber beating at 21,600 vph with a power reserve of 40 hours. The silvered guilloché dial showcases a tourbillon with small seconds on the cage at 12 as well as an astronomic moonphase at 6. The silvery, cool color of the dial is complemented by the 18K white-gold case, which is perfectly set with two rows of diamonds (1.4 carats). This model possesses two of the grand complications all while exhibiting elegance and beauty.

TAG HEUER

MONACO V4 – REF. WAW2080.FC6288

The Monaco V4 has always been an extremely technically sophisticated watch, as it boasts an automatic belt-driven transmission, a tungsten ingot linear mass, and two pairs of barrels in series, set in parallel and linked by belts. The latest model, crafted in titanium, ruthenium and ceramic, showcases TAG Heuer's masterful use of the high-tech materials. The grade-5 titanium case now possesses a 41mm diameter, and fine-brushed "ceramic" inserts visually lighten the watch's squared shape. Other titanium components are the crown and the strap's folding buckle. The V4 movement's avant-garde architecture remains unchanged, but the finishes now incorporate a high-resistance black ruthenium coating on the back plate and the Côtes de Genève-decorated bridges. This timepiece is available in a limited edition of 200 pieces.

VACHERON CONSTANTIN

PATRIMONY CONTEMPORAINE PERPETUAL CALENDAR – REF. 43175/000R-9687

On its silver-toned opaline dial with baton- and triangle-shaped 18K pink-gold hour markers, this elegant timepiece presents a perpetual calendar so effortless that the moonphase at 6:00 seems to conduct the entire time orchestra. Via a month dial at 12:00, Vacheron Constantin provides an optimally unambiguous display of the leap-year indicator residing in its interior. Contained within the 41mm 18K pink-gold case, the Poinçon de Genève-honored 112 QP self-winding movement beats at a frequency of 19,800 vph, with a 40-hour power reserve. Turned over, the openworked back can be admired through a sapphire crystal.

VACHERON CONSTANTIN

PATRIMONY TRADITIONNELLE 14-DAY TOURBILLON – REF. 89000/000R-9655

An exceptional tourbillon and the absolute purity of a round, understated case housing a complication boasting unprecedented performances: the Vacheron Constantin Calibre 2260 beating at the heart of the Patrimony Traditionnelle 14-day Tourbillon is simply unequalled. In order to provide a full 14-day power reserve, the caliber is equipped with four barrels mounted in coupled pairs. Moreover, the latest newcomer to the Patrimony Traditionnelle line was the first Vacheron Constantin watch to be approved according to the updated Poinçon de Genève criteria; the certification no longer concerns the movement alone, but rather the timepiece as a whole. All of these remarkable components combined make this model a predestined collector's item.

ZENITH

ACADEMY CHRISTOPHE COLOMB HURRICANE – REF. 18.2210.8805/01.C713

The Academy Christophe Colomb Hurricane has no equivalent in the field of haute horology. With its gyroscopic module that enables the regulating organ to remain constantly horizontal—however the watch itself is positioned—and thus to protect the mechanism from the effect of earthly gravity, the line of Christophe Colomb models is truly unique. This principle ensures enhanced regularity of the escapement, without the position of the watch interfering with the rate of the movement—an approach that thereby improves precision. By also incorporating a fusée and chain escapement that regulates the torque transmitted to the gear trains, the Christophe Colomb Hurricane model offers an impressive combination of advanced features.

Equations of Time

Jaeger-LeCoultre
Gyrotourbillon 1

François-Henry Bennahmias

CEO of Audemars Piguet

"The Royal Oak appeals to every generation"

48-year-old François-Henry Bennahmias was appointed CEO of Audemars Piguet in May 2012. Having joined the company in 1996, he served from 1997 onwards as Assistant Manager, Marketing & Sales for Asia-Pacific and Europe. In 1999, he became CEO of Audemars Piguet North America Inc. in New York. **HIS COMMITMENT, HIS NETWORK AND HIS DYNAMISM** have enabled him to successfully develop the brand in the United States. Now located at the company headquarters in Le Brassus, he agreed to share some of his plans and expectations for the coming years.

ROYAL OAK OFFSHORE GRANDE COMPLICATION

If you were to explain to a neophyte what Audemars Piguet stands for, what would you say?

It is first and foremost a large family business, which has left an indelible imprint on watchmaking history. In its 138 years of existence, it has set countless records and introduced numerous innovations and extraordinary watches that have all contributed to forging a unique and instantly recognizable style. Audemars Piguet is a "never-ending story." The brand's characteristic creativity, expressed through all aspects of its activity, is the very essence of its longevity.

What makes Audemars Piguet different from other watch brands?

The art of surprising observers involves a lot of hard work. Since its founding in 1875, Audemars Piguet has consistently striven to offer a blend of excellence and novelty. This experience now enables us to venture off the beaten track and to pop up where we are least expected. It is this great strength that enabled us in 1972 to launch the Royal Oak, which is still our best-selling model.

2012 was the Royal Oak's 40th birthday: what is your most vivid memory of this anniversary year?

Public recognition and appreciation! This icon has an extraordinary ability to make its way through the decades unscathed. 40 years on, the Royal Oak is in fact 40 years young! It is doubtless the only fine watchmaking model on the market that attracts 20-year-olds as much as older folks, as well as both men and women across Europe, America and Asia. It truly has a contagious and universal appeal. I indeed observed the particular interest shown by young people at the various exhibitions held around the world to celebrate this anniversary.

Does this specific Royal Oak recognition carry over to the brand itself?

Definitely! The Manufacture, the Vallée de Joux and Swiss expertise have always been an integral part of the communication about the Royal Oak. Through this model, our customers had an opportunity to become more intimately acquainted with all that. The prestige of the watch is inextricably bound up with that of the House itself.

Which complication fascinates you most in the current Audemars Piguet range?

The Equation of Time. First because it is has a significant history at Audemars Piguet: the first pocket watch equipped with an equation of time emerged from the company workshops before the turn of the century; in 1925 we launched the first model with a perpetual calendar lined up along the 12 – 6 o'clock axis; and in 1928 introduced a perpetual calendar pocket watch featuring a "running" equation of time hand. But what touches me most in this complication is the encounter between poetic elegance and technical sophistication. The Jules Audemars Equation of Time displays a complete perpetual calendar, sunrise and sunset times, as well as an astronomical moonphase display. As far as the two latter functions are concerned, this highly complex movement can be adjusted to the city of one's choice: the ultimate intensely personalized and truly romantic touch.

How is Audemars Piguet's current range divided between men's and ladies' watches, and between quartz and mechanical models?

Around 70% of our sales come from men's watches. As far as quartz watches are concerned, we make hardly any of them these days.

And what proportion does the Royal Oak represent?

About 65% of our sales.

ROYAL OAK EQUATION OF TIME

Isn't that a bit lopsided?

You won't see us complaining about the phenomenal success of the Royal Oak! We do of course aim to strike a balance between our collections, but that won't be done by reducing or putting a brake on the Royal Oak—on the contrary! This model enjoys continuous success and enables us to grow. We don't plan to change anything as far as that's concerned. The balance will be achieved by pushing the other collections.

Did the international context have an adverse effect on the 2012 results?

We had an excellent year in 2012! Nonetheless, the poor state of the worldwide economy naturally had an impact on people's thinking. However, we are fortunate enough to have fairly evenly spread markets. And even though Asia has been experiencing a slight drop-off, one should not generalize: Singapore is doing very well for example, as is Japan, which remained stable and even saw a slight increase in sales. Meanwhile, it was a very good year in the United States.

Which are Audemars Piguet's main markets?

We are present around the world in a very balanced way, with a representation that is well above average in the Americas and slightly below that of some others in Europe and Asia.

You have recently taken the helm of Audemars Piguet. What are your priorities and the key upcoming projects?

Within a tough economic setting, we have decided to consolidate the foundations of our brand. We do not wish to increase production, but on the contrary to work on quality—not only as far as the product is concerned, but on all levels. This involves improving relations with our biggest retailers and strengthening training programs both in watchmaking and sales professions. In Audemars Piguet movements, even the parts that remain hidden must be perfect! I intend to extend this concept to the entire company. We aim for excellence in all areas, including production, sales and services.

ROYAL OAK TOURBILLON CHRONOGRAPH

ROYAL OAK GRAND COMPLICATION

Is that part of a global vision of luxury?
In all modesty, I would say it is.

And is there a lot left to do?
There is always plenty to be done! We are in such a complex industry where the human factor intervenes in such a wealth of details that there is always something worth improving. The goal is to finish fine-tuning the last details with a view to strong redeployment in 2015.

On a global level, how would you say the brand has evolved over the past five years?
We have just completed a major consolidation phase: we have bought up our distributors one after the other and we now enjoy almost 90% control of our markets. This enables us to establish and convey global messages and to project a unified image.

Is the brand perceived differently on various continents?
We are indeed doing everything in our power to avoid that—but I think we still have work to do and that notably involves distribution. Meanwhile, we have also internalized a wide range of skills within the company, the latest being dial-making. Around 2,000 magnificent dials were already made in 2012! This was no easy challenge. We are thus determined to achieve maximum vertical integration in terms of both our production and our distribution.

What is your policy as far as own-name boutiques are concerned?
We have around 20 and will be gradually increasing that number—but that's a trend typical of the entire watch industry today.

Does that spell the end of independent retailers?
Absolute not—on the contrary. We know that we didn't reach our current heights under our own steam! But we absolutely must work with people prepared to challenge their working methods into question and prepared to make progress.

And what effects are the Internet and social media having in the field of selling luxury goods?
This also forces us to take a fresh look at our approach, for this is a new communications territory and a new way of thinking that we simply cannot afford to ignore. Some industries have a head start on us in this field, but I think we're not doing too badly—and the considerable leverage effect already being exercised is certainly impressive.

ROYAL OAK CONCEPT CARBONE

The “equation of time” display—

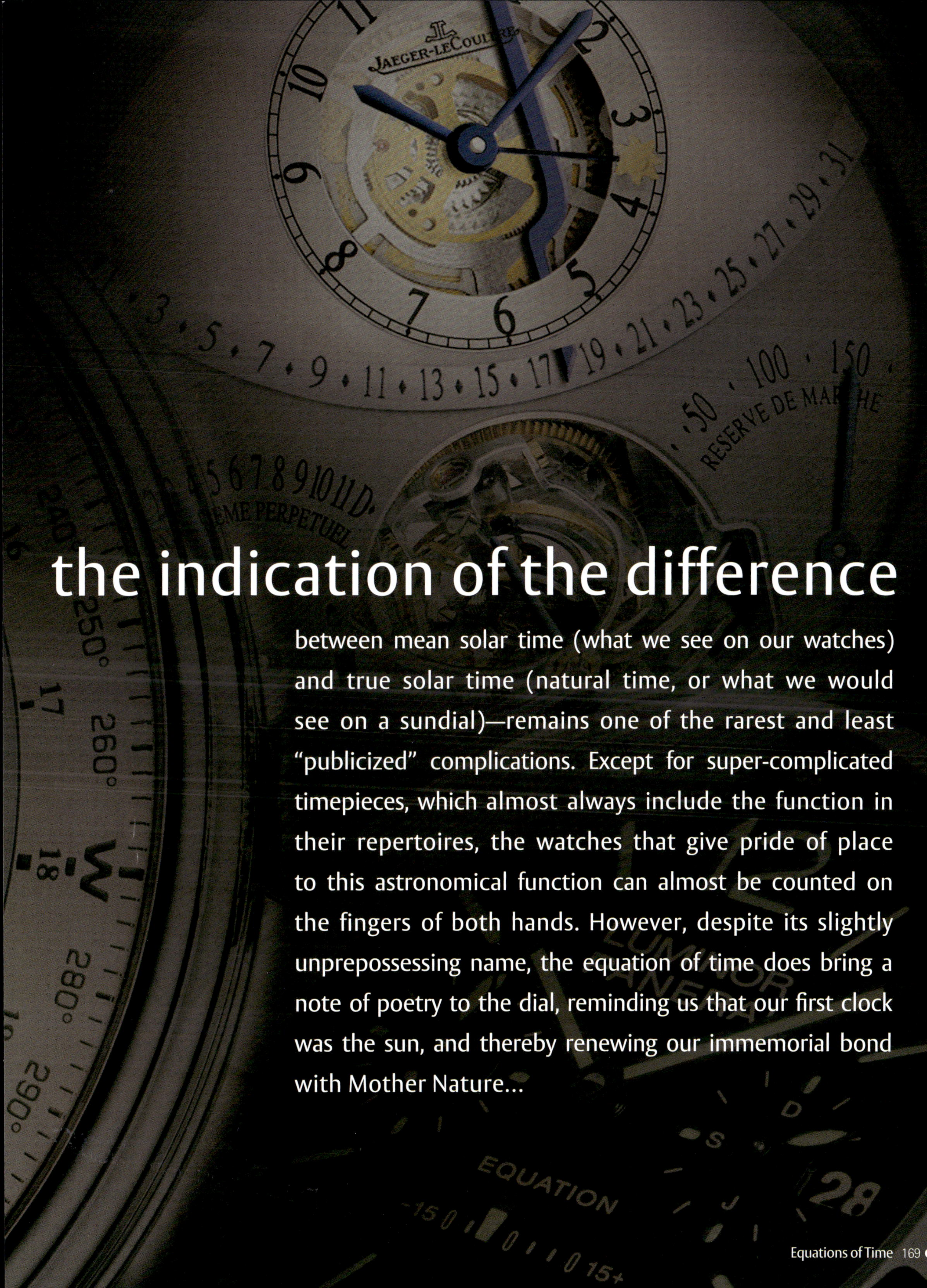

the indication of the difference

between mean solar time (what we see on our watches) and true solar time (natural time, or what we would see on a sundial)—remains one of the rarest and least "publicized" complications. Except for super-complicated timepieces, which almost always include the function in their repertoires, the watches that give pride of place to this astronomical function can almost be counted on the fingers of both hands. However, despite its slightly unprepossessing name, the equation of time does bring a note of poetry to the dial, reminding us that our first clock was the sun, and thereby renewing our immemorial bond with Mother Nature...

There are actually two times: true solar time and mean solar time. True solar time, which you might see on a sundial, corresponds to natural rhythms. The Earth describes an ellipse in turning around the sun, and its axis is inclined in relation to its orbit. This means that the lapse of time between two passages of the sun through its highest point in the sky (noon) is not the same length year-round. It actually lasts 24 hours only four times a year: on April 15th, June 14th, September 1st and December 24th. Otherwise, it is sometimes longer and sometimes shorter, following an immutable curve. This difference ranges from 16 minutes and 23 seconds, on November 4th, to 12 minutes and 22 seconds, on February 11th and is called the "equation of time." As for mean solar time (also called civil time), it is a convention based on the average of all days in a year.

Since the famous Astrarium clock by Giovanni de' Dondi (1364), the greatest watchmakers have rivaled each other in ingenuity to develop systems that reproduce the variations in the equation of time, first in clocks, then, since the turn of the 18th century, in wristwatches. This function has become an indispensable feature of all ultra-complicated watches—particularly the astronomical ones—but it is much more rare to find it flying solo. In 2004, the "running" equation of time (see "How does it work?"), long reserved for pocket watches, made its wristwatch debut.

TECHNIQUE

HOW DOES IT WORK?

Since the variable lengths of true solar days are identically repeated on the same days, they can be "programmed" by means of a so-called "equation" cam that performs one full rotation per year. Its profile resembles that of an "analemma," representing the figure eight traced in the sky by the sun's various positions as recorded each day at the same time and from the same place throughout the calendar year. The extreme precision of this shape determines the accuracy of the equation of time.

DISPLAY METHODS

The equation of time display may be presented in various ways. Most watches are equipped with a subdial or auxiliary segment swept over by a hand running from -16 to + 14 minutes (sometimes from -15 to +15); to know true time, the user must then do a mental addition or subtraction of this difference in relation to mean time.

RUNNING EQUATIONS

Distinctly more user-friendly yet far more complex in terms of their construction, "running" equation of time watches have two hands fitted coaxially with the minute hand, respectively indicating mean time and true time. They are equipped with a sensor system that searches out the information from the cam, as well as a differential that corrects the running of the supplementary minute hand. They require no mental arithmetic, since true time is continuously displayed directly on the dial.

BLANCPAIN
JAN
MAR
MAY
JUL
SEP
NOV
L
1
2
3
+15
0
-15
SUN
MON
TUE
WED
THU

A Very Rare Complication

Extremely rare as a solo complication, the equation of time is most often joined by a perpetual calendar, even an annual calendar, or with astronomical indications. It is usually displayed on a small fan-shaped segment covered by a hand, but certain watchmakers prefer to innovate by spotlighting the complication with a central hand or opting for a linear indication.

Chopard, *L.U.C 150 All-in-One*

Chopard uses the back of its anniversary model L.U.C 150 All-in-One as a second dial: it features a power reserve display, 24-hour night and day indication, sunrise and sunset times, an orbital moonphase and an equation of time—all set to correspond to what may be observed in Geneva.

Panerai

Luminor 1950 Equation of time
Tourbillon Titanio 50 mm L'Astronomo

The Luminor 1950 Equation of time Tourbillon Titanio 50 mm L'Astronomo provides an innovative linear display of the equation of time at 6 o'clock. It also indicates the times of sunrise and sunset for a reference city chosen by the client, and its caseback bears a depiction of the night sky as seen from that same city.

Zenith
Academy Christophe Colomb Equation of Time

Equipped with an escapement that maintains a horizontal position, whatever the position of the watch—as with the gimbals used in maritime navigation—the Academy Christophe Colomb Equation of Time from Zenith displays the equation of time across a fan-shaped segment at 9 o'clock.

Girard-Perregaux

1966 Annual Calendar and Equation of Time

The 1966 Annual Calendar and Equation of Time by Girard-Perregaux is distinguished by its dial, on which an extremely original configuration features the date in a small subdial at 1:30, the small seconds at 9 o'clock, the month in a panoramic window at 7 o'clock and the equation of time on the arc of a circle at 4:30.

Breguet

Classique Complications 3477BA/1E/986

On its finely guilloché dial, this grand complication watch from Breguet's Classique collection combines a perpetual calendar, power reserve display and an equation of time indication in a fan shape at 1:30.

Vacheron Constantin

Patrimony Traditionnelle Calibre 2253 Collection Excellence Platine

The Vacheron Constantin Patrimony Traditionnelle Calibre 2253 Collection Excellence Platine is endowed with a tourbillon movement powering indications of the perpetual calendar and times of sunrise and sunset, as well as the equation of time on the arc of a circle at 10:30.

Girard-Perregaux

1945 Vintage Perpetual Calendar and Equation of Time

Housed in a rectangular Art Deco-inspired case, the 1945 Vintage Perpetual Calendar and Equation of Time from Girard-Perregaux has been released in a version with a wperpetual calendar and an equation of time segment-type display appears in the upper left corner of the dial.

Daniel Roth, *Perpetual Calendar Equation of Time*

Pride of the Masters Grandes Complications collection, the Perpetual Calendar Equation of Time model from Daniel Roth boasts an open-worked dial that allows glimpses of the movement as well as the different calendar discs.

Taking Longitude into Account

By convention, all the places situated in one time zone have the same time. But "true" solar noon varies by longitude, and it is distinctly earlier in Warsaw than in Madrid. Executing a correction of -16 to +14 minutes over an average applicable to the entire time zone (60 minutes) is therefore not of any great scientific interest, but innovative watchmakers are remedying this flaw with models that take into account the exact degree of longitude.

Audemars Piguet, *Royal Oak Equation of Time*
The first brand to have developed an equation of time indication system adjusted to the exact degree of longitude, Audemars Piguet equipped its Royal Oak Equation of Time with a large central hand for the complication's display. The inner bezel ring bears the name of the reference city chosen by the client, as well as the time of true noon in that city. When the hour hand corresponds to the hour indicated on the inner bezel ring, and the equation of time and minute hands line up exactly, the owner of the watch knows that the sun is exactly at its zenith. This is also one of the rare wristwatches to display sunrise and sunset times while taking into account three parameters: date, longitude and latitude.

Arnold & Son, *True North Perpetual*
The True North Perpetual watch from Arnold & Son takes account of longitude, thanks to a dial featuring two mobile rings. The model features another unprecedented characteristic: when the 24-hour solar hand indicates the true solar noon, the wearer need only point it towards the sun for the mobile outer dial to indicate true geographical North—hence its name.

Running Equations of Time

Distinctly more user-friendly yet far more complex in terms of their construction, "running" equation of time watches have two hands fitted coaxially with the minute hand, respectively indicating mean time and true time. They require no mental arithmetic, since true time is continuously displayed directly on the dial. This system, which had long been regarded as the exclusive preserve of pocket watches, first appeared on wristwatches in 2004 and represents a genuine feat in terms of miniaturization.

Blancpain
Villeret Equation Marchante Pure

Blancpain has associated a running equation of time display and a perpetual calendar in a model from the Villeret collection named Equation Marchante Pure. A circular opening at 6 o'clock allows the wearer to contemplate the cam that governs the equation, as well as the sensor in action.

Jaeger-LeCoultre
Gyrotourbillon 1

The Gyrotourbillon 1 by Jaeger-LeCoultre, equipped with a spherical tourbillon (see chapter on Tourbillons), also features an offset hour/minute subdial bearing a running equation of time displayed by an additional minute hand adorned with a tiny golden sun.

Blancpain
Le Brassus Equation Marchante

The Le Brassus Equation Marchante by Blancpain is distinguished by its double equation of time display: the running equation of time itself appears in the center with an additional central minute hand tipped with a golden sun; and a subdial at 2 o'clock boasts a retrograde hand indicating the number of minutes of difference. This sophisticated device is combined with an ultra-thin perpetual calendar.

AUDEMARS PIGUET

ROYAL OAK EQUATION OF TIME – REF. 26603ST.OO.D092CR.02

The Royal Oak Equation of Time is housed in a stainless steel case and powered by an automatic-winding 2120/2808 movement, which beats at 19,800 vph and is endowed with a 42-hour power reserve. Visible through the sapphire crystal caseback, the 41-jewel movement depicts conventional time and real solar time and offers hours and minutes, sunrise and sunset times, equation of time, perpetual calendar and moonphase. The solar zenith is calibrated to the reference city chosen by the owner. The openworked oscillating weight may be personalized on request (the standard version has the "AP" logo). Mounted on a large scale hand-sewn crocodile leather with AP folding clasp in stainless steel, this model is also available in 18K pink gold.

BLANCPAIN

VILLERET EQUATION DE TEMPS MARCHANTE – REF. 6638-3631-55B

The Equation Marchante Pure, based on the Le Brassus model that was the first wristwatch capable of displaying the separate solar and civil times by two minute hands, is an integral part of the Villeret collection, characterized by its pure lines and formal beauty. Fitted with a three-subdial perpetual calendar complete with a retrograde moonphase, this red-gold masterpiece delights with an aperture at 6:00, through which the owner is treated to the precise analemma-shaped equation of time cam. This breathtaking piece houses the 364-part caliber 3863A, equipped with 72 hours of power reserve.

VACHERON CONSTANTIN

PATRIMONY TRADITIONNELLE CALIBRE 2253 – REF. 88172/000P-9495

A hand-stitched dark blue alligator strap with platinum threads traces the path to a truly astonishing spectacle. Limited to 10 numbered pieces, this 44mm 950 platinum timepiece houses a remarkable manually wound caliber honored with the Poinçon de Genève. While the perfectly finished tourbillon with small seconds at 6:00 completes its rotation in precisely one minute, the rest of the dial looks at time on a significantly grander scale. Set to the owner's reference city of choice, the sunrise and sunset indicators surround the exposed carriage on either side. On the upper half of the watch face, the perpetual calendar, complete with day of the week, month, date, and leap-year indicator, is joined by the equation of time between 10:00 and 11:00. Merging the tourbillon's symbolism of horological history with a wealth of highly sophisticated functions, the Patrimony Traditionnelle Calibre 2253 makes a magnificent statement regarding the grandeur of time.

ZENITH

ACADEMY CHRISTOPHE COLOMB EQUATION OF TIME – REF. 18.2220.8808/01.C631

The 2011 prize in the stellar "Best Complicated Watch" category was won by the Academy Christophe Colomb Equation of Time. The jury of the Geneva Watchmaking Grand Prix thereby rewarded a three-fold technical achievement: liberation from gravity, high frequency and visualization of the duration of the Earth's rotation, which is not an immutable 24 hours, since the Earth is in fact from day to day either ahead or behind the time shown on a watch.

Tourbillons

Franck Muller
Giga Tourbillon

"Each De Bethune watch meets three specific criteria: tradition, innovation and design."

A FULL-FLEDGED MANUFACTURE

founded in 2002, De Bethune is now clearly one of the watch brands capable of bringing fresh momentum to the field of time measurement. Its fundamental principles are conveyed through models that are both imbued with tradition and yet resolutely innovative—two values that are in turn governed by an instantly recognizable aesthetic approach.

De Bethune has got the wind in its sails. With 340 watches sold in 2012, as brand CEO Pierre Jacques explained, the brand has more than tripled sales in two years and shown a profit for the past three. "In a nutshell," said Jacques, "De Bethune may be described as a small independent manufacture with a 60-strong personnel, which performs all its production in-house—from the movements to the dials, cases and hands." The CEO also underscored the role played by the "Aiguille d'or" Grand Prix—the supreme award bestowed by the Geneva Watchmaking Grand Prix (GPHG) and which the brand won in 2011 with its DB 28 Titanium. "This Prize clearly took us to a new level," he commented. "The trust that customers and suppliers had shown us until that point was strengthened and the small De Bethune found its place on the international watchmaking map. It's a kind of winning wager on the future."

DE BETHUNE
Manufacture

DAVID ZANETTA
President

PIERRE JACQUES
CEO

DENIS FLAGEOLLET
Technical Director

THREE DESIGN CRITERIA

DB16

DB25 IX MAYAN UNDERWORLD

One must indeed admit that the GPHG jury showed definite flair in honoring this model in 2011. The DB 28 is an exceptional timepiece both technically and aesthetically. The design is in line with the brand DNA, expressed through the specific shape of its case (which bears the crown at 12 o'clock as a nod to pocket watch tradition) and its spherical moonphase display at 6 o'clock. The DB2115 caliber powering it is equipped with a self-regulating twin barrel and a silicon/palladium balance inside the regulating organ with its flat end curve protected by the triple pare-chute shock-absorbing system. Commenting on this award at the prize-giving ceremony, David Zanetta, President and co-founder of the brand along with Technical Director Denis Flageollet, pointed out: "The idea behind De Bethune is that horology is both an art and a science. For us, it is about giving a new lease on life to historical watchmaking with a view to the future of humankind. There is thus no question of doing more, but instead of doing better."

In this respect, Jacques specified: "Each De Bethune watch meets three specific criteria: tradition, innovation and design. That means that we make our timepieces in keeping with the rules of the art, including the finishing; that we equip them with the latest technologies and the most high-performance materials; and that their instantly recognizable architecture reinterprets shapes from the past in terms of a contemporary vision. One can thus truly speak of a De Bethune style." This style is uncompromising when it comes to development. Under the aegis of Flageollet, the brand has set up a ten-strong R&D department that has a large number of inventions to its credit, notably in the field of silicon. Summing it up in hard numbers: in its ten years of existence, the brand has filed nine patents, presented 17 world-firsts and created 15 in-house calibers, of which the most

recent—named S233 in reference to the specific gravity of silicon—is a De Bethune base movement that may be enhanced with all known horological complications and already powers the DB 27 Titan Hawk.

VENTURING INTO THE FIELD OF HIGH FREQUENCY

In David Zanetta's view, the most recent authentic horological innovations date back to the 1950s, before the advent of quartz compromised this inventiveness. In this respect, fundamental research must play its rightful role in order to revive the grand watchmaking traditions of the past, based on mechanical and mathematical concepts and supported by the work of masterful artisans. With this in mind, De Bethune has adopted an approach that places it firmly in the vanguard of the "concept watch" field, particularly in seeking to break free of the traditional lever escapement. The solution being explored involves synchronizing a small magnetic-escapement rotor and an acoustic resonator. As far as distribution is concerned, there are already signs of achieving mastery of frequencies ranging between 100 and 200Hz (720,000 – 1.44 million vph), and potentially even 1,000Hz.

That does not mean De Bethune neglects the artistic crafts, as was confirmed by the DB 25 "IXème inframonde Maya," a model with a gold dial inspired by the arithmetic of Mayan time and hand-engraved by Michèle Rothen to stunning effect. This combination, characteristic of the brand, which plays with various horological complications with astounding facility while highlighting the skills of the finest artisans, is earning ever-greater appreciation. At the end of 2012, De Bethune became part of the Fondation de la Haute Horlogerie (FHH), thereby, according to the latter, joining the circle of "firms wishing to promote, across all five continents, a form of watchmaking founded on centuries of tradition and cutting-edge technology." This new partnership came hot on the heels of the extension of the manufacture located in the village of L'Auberson in the Vaudois Jura—once again demonstrating that De Bethune firmly intends to focus on equity investments in its production facilities.

STRENGTHENED DISTRIBUTION

Extension is also the name of the game on its distribution network. "The brand is now represented in around 30 points of sale on the main international markets via partnerships with well-reputed distributors with a perfect understanding of the De Bethune philosophy," said Jacques. "Not to mention its own-name boutique in Geneva, which is an excellent idea in terms of brand image, as well as commercially. Within the next two years, we are aiming for around 40 points of sale for an annual production of around 500 watches." This is a perfectly realistic objective, especially as 2013 will see De Bethune produce grand complication watches, mysterious table clocks, dials featuring a star-studded sky, a new chronograph... In short, plenty to keep the flame of horological passion burning!

SKYBRIDGE

François-Paul Journe

Founder of F.P. Journe

"For a watch, courtesy means precision"

THE NEW MASTERPIECE in the Souveraine collection, the Chronomètre Optimum, is the latest creation from F.P. Journe. Everything about its conception, from the power reserve to the escapement as well as the balance and the remontoire or constant-force device, has been designed to achieve just what its name implies: optimal precision. François-Paul Journe, who developed this project for over 12 years, shares some explanatory details.

Montres Journe was founded in 1999. How many people does it now employ?
Around 150, including the boutique staff.

Is that the ideal size?
I'm not planning to grow much more. There are currently 25 watchmakers, a number that may grow to 30 in the coming years—not to increase production, but in step with other projects. Within the watch industry, growth at all costs leads nowhere.

What is your current production volume?
Around 850 watches a year.

You are developing an own-name boutique network around the world. How many of them are there to date?
The latest opening in Miami brings us to eight in all! Each is an adventure in its own right, since it takes a lot of ingredients to make a boutique: an ideal location, reasonable operating expenses, qualified staff... I don't have all the clout of a big group! But I do believe that if we hadn't developed our boutique network, we would no longer be around today.

What share of your sales do your boutiques account for?
Around half, not counting the new ones. My aim is to reach 65% to 70%. I think the real problem for independent brands in the future will be sales.

And what about the traditional points of sale?
We have around 30 of them, but we are not looking for other partners—simply because we couldn't keep them supplied.

TOURBILLON SOUVERAIN

One of your latest accomplishments is the Chronomètre Optimum. What was the approach behind this model?
Almost 25 years ago, I told one my collector friends that if I were to make a watch for myself, I would do something simple with three hands, a power reserve ensured by a twin barrel, a remontoire or constant-force device, and a special escapement. So it was something I already had in mind and that I had even sketched in 2001, without ever having time to actually pursue the project.

The Chronomètre Optimum is indeed equipped with a remontoire. What exactly does that mean?
I made my first wristwatch in 1991. It was a tourbillon. The problem with this horological complication is that when the mainspring force dwindles, the frictional forces on the tourbillon increase proportionally and this leads to a far greater loss in amplitude of the balance than with a classic movement. I thus had the idea of adding a remontoire, which is a system fitted just before the escapement and which provides the movement with a supply of constant driving force. I decided to use this mechanism on the Chronomètre Optimum, while further improving it to enhance precision.

You are among those who have reminded us that precision could be important for a watch...
I like to say that for a watch, courtesy means displaying the exact time!

What kind of improvements are you referring to?
Since 2001, we have been pushing the vertical integration of the company and made some significant advances in the technical features of our watches. Three or four years ago when I went back to this project, I requested that it should be built from scratch so as to incorporate the latest methods and contemporary design. As far as the remontoire is concerned, we enjoyed the benefit of a certain hindsight...

Are you the kind of watchmaker who says that movements are never definitively complete, but are instead in a perpetual state of evolution?
Definitely! On the 1991 tourbillon for example, the remontoire was not properly balanced. Depending on the positions or on wrist movements, there could be slight differences in the transmission of the driving force. While these differences were minimal, if we were working on a piece called Optimum, we had to do everything in our power to ensure it lived up to its name! Its remontoire is thus finely balanced and is also made from titanium to make it lighter. The resulting constant force guarantees isochronism for 45 hours. That means that over any arc of amplitude, the balance performs its to-and-fro motion over exactly the same period.

Which are this model's other key characteristics?
There are two parallel-mounted barrels that operate with the first wheel, the great wheel, which is almost between the two. This arbor is almost free of any lateral pressure, which is important. The power reserve than operates with the one-second remontoire. It was at that point that, in designing this movement, I realized that there was a wheel that quite naturally revolved in 60 seconds—which is why I thus displaced the arbor and fitted it with a dead beat seconds.

What about the escapement?

I had been wanting to make a lubricant-free escapement for some time. It has always been one of my motivations, especially since watchmaking is well acquainted with oil-free escapements thanks to marine chronometers. But those are what we call detent escapements, of the direct-impulse type. In a lever escapement, the oil is extremely volatile because it has no place to adhere. While considerable progress has been made in recent years, movements still require fresh lubrication after three to four years.

So what was your solution?

These days, watchmakers wishing to make lubricant-free escapements use silicon, which enables dry lubrication. I have worked with silicon, but I found it too fragile to be used in a luxury product such as ours. In 50 years, if it were to require repair, any such intervention would have to be conducted in a specialized laboratory! I therefore did not opt for this path, but instead tried to apply DLC (Editor's note: Diamond-Like Carbon, a coating with an extremely low friction coefficient achieved by physical vapor deposition (PVD) and combining the structure of graphite-like carbon with that of diamond) on more solid metals. We were fairly successful, but we never managed to stabilize the system—probably because the companies performing these processes are not used to working with tiny watchmaking components. Finally, I equipped the Chronomètre Optimum with a direct-impulse escapement made from traditional materials—which means any p otential repairs can be done by a good watchmaker.

What is special about this escapement?

It is a Bi-Axial High Performance (BHP) escapement. I devised a system deriving the best aspects of a direct-impulse escapement on a marine chronometer with the advantages of a wristwatch lever escapement. With a double direct impulse, friction is reduced by half because there are two impulses instead of one for the same amount of energy. Meanwhile, the balance-spring is fitted with a Philips end curve, which enables concentric development of the spring, thereby avoiding any displacement of the center of gravity. Finally, a perfectly symmetrical and balanced lever ensures virtually no loss of amplitude.

So you delved back into watchmaking history to track down the best features of the various escapement techniques?

Watch screws and wheels were invented several hundred years ago! We naturally draw on history when inventing new things. I feel I am very much part of this momentum that has been driving horology forward since the 17th century and the invention of the balance-spring. The watch belonging to Louis XIV had just one hand, the minute hand appeared after the invention of the balance-spring, and the seconds hand only emerged in the latter part of the 18th century. Almost one hundred years went by between the minute hand and the seconds hand. However, for the latter to serve any useful purpose, precision had to be improved—and that is still my objective today.

Do you plan to equip all your models with this escapement?

No, we need a little bit of distance to apply this globally. But I do plan to have the Chronomètre Optimum compete in the next Chronometry contest.

How do things currently stand in terms of precision?

At around plus or minus one second per day—although that obviously depends on how the watch is worn.

CHRONOMETRE OPTIMUM

Richard Mille

CEO of Richard Mille

"The brand is like a two-stage rocket"

Beneath the outward impression of **CONSTANTLY PUSHING THE ENVELOPE**, Richard Mille is in fact carefully orchestrating the brand's progress like a perfect watch mechanism, while demonstrating an ability to continue inspiring and nurturing dreams among collectors as well as younger generations. This is achieved by a subtle chemistry between caution and innovation.

Just as at the end of every year, Richard Mille was in a huge rush at the time of this interview in late 2012. Delivering the models presented at the start of the year is a daunting challenge, due to all manner of delays across the board on the collections that hamper the imperative need to honor orders. "At the end of the day, that's what determines our turnover," explained the watchmaker, fairly serenely, especially when one considers that the brand experienced around 20% growth in sales amounting to around 110 million CHF for 2,700 watches.

RM 039 TOURBILLON FLYBACK AVIATION

Why do you always end up facing the same end-of-year problem?

It is due to the fact that our watches are anything but simple, and that any smooth production flow concepts are totally illusory as far as we are concerned. In other words, when we start from scratch in creating a model of which only around 20 will be produced—as is often the case with Richard Mille watches—we are often faced with certain supply difficulties in terms of delivery deadlines. Such is for example the case with the Tourbillon RM 039 Aviation E6-B Flyback, which comprises almost 1,000 parts and with which we are running about a year and a half late for a delivery scheduled for the end of 2013. I am nonetheless pleased to note that Renaud & Papi (editor's note: owned by Audemars Piguet), which is handling the development, considers this chronograph to be the most complete ever made.

Might one say that Richard Mille is a victim of its own success?

I don't think you can really put it those terms, but it is certainly true that our limited series are extremely successful. The RM 056 Felipe Massa Sapphire is a good example. Its three-part case, made entirely from sapphire crystal, definitely appealed to our customers. On the evening it was first presented, the five-piece series was sold out. We could even have sold five times more. I will thus now work with

The most exclusive segment—which is only costly in terms of "brain power" in that everything is sold before production even begins—enables us to build a fantastic body of knowledge that then serves the brand as a whole. From the first prototype of the "Nadal" watch to the final version, we broke five or six watches, and one inevitably draws lessons from such incidents. The same goes for the RM 031 High Performance, for which the main goal was to create an extremely accurate watch with a variation in rate of less than 30 seconds per month. At this stage of our learning curve, we have reached the 15- to 20-second range. So yes, we do sometimes run late and the development of some of our products does indeed sometimes take longer than expected. But what counts is that we are all motivated by the idea of innovating and that we have fun.

Contagious fun?

According to our customers, we have indeed managed to create a certain appreciation of our approach by working on horological innovation in extreme conditions. The brand thus exercises a cross-generational appeal among enthusiasts, whether they are collectors or part of the up-and-coming generations. I like to say that we have earned our credibility on the battlefield. Each watch brand has its own philosophy. Ours has inspired dreams and enabled us to carve out a niche. That is what drives us to cultivate the creative consistency I mentioned earlier. Establishing a parallel with the automotive industry, nobody points the finger at car manufacturers that simultaneously produce sedans, sportsters, SUVs and minivans. If a brand has plenty of creativity, it might as well express it!

What highlights can we expect in 2013?

Alongside the hundred or so points of sale where we are already represented, we are planning to extend our boutique network, again in partnership with local distributors. There are many countries where we are not present, such as Germany and Korea. We also want to strengthen our presence in the United States. On the product side, we are continuing to serve our partners, notably with a new "Nadal" watch, with the model developed for Yohan Blake, double medalist at the London Olympics, and the upcoming one for Sebastien Loeb, the nine-time world rally driving champion who has just joined the Richard Mille team. We are also presenting a Panda watch and two Jean Todt models: a Worldtimer and the Tourbillon G-Sensor RM 036, designed to mechanically and visually convey the number of Gs accumulated by the wearer during rapid decelerations. 2013 will also be a very important year for us because we are inaugurating a second case-making factory, thereby securing perfect independence in this field, apart from sapphire cases. We are determined to push the brand's vertical integration in order to meet our own needs. In a nutshell, this will be yet another year full of thrills and surprises.

RM 59-01 TOURBILLON
YOHAN BLAKE

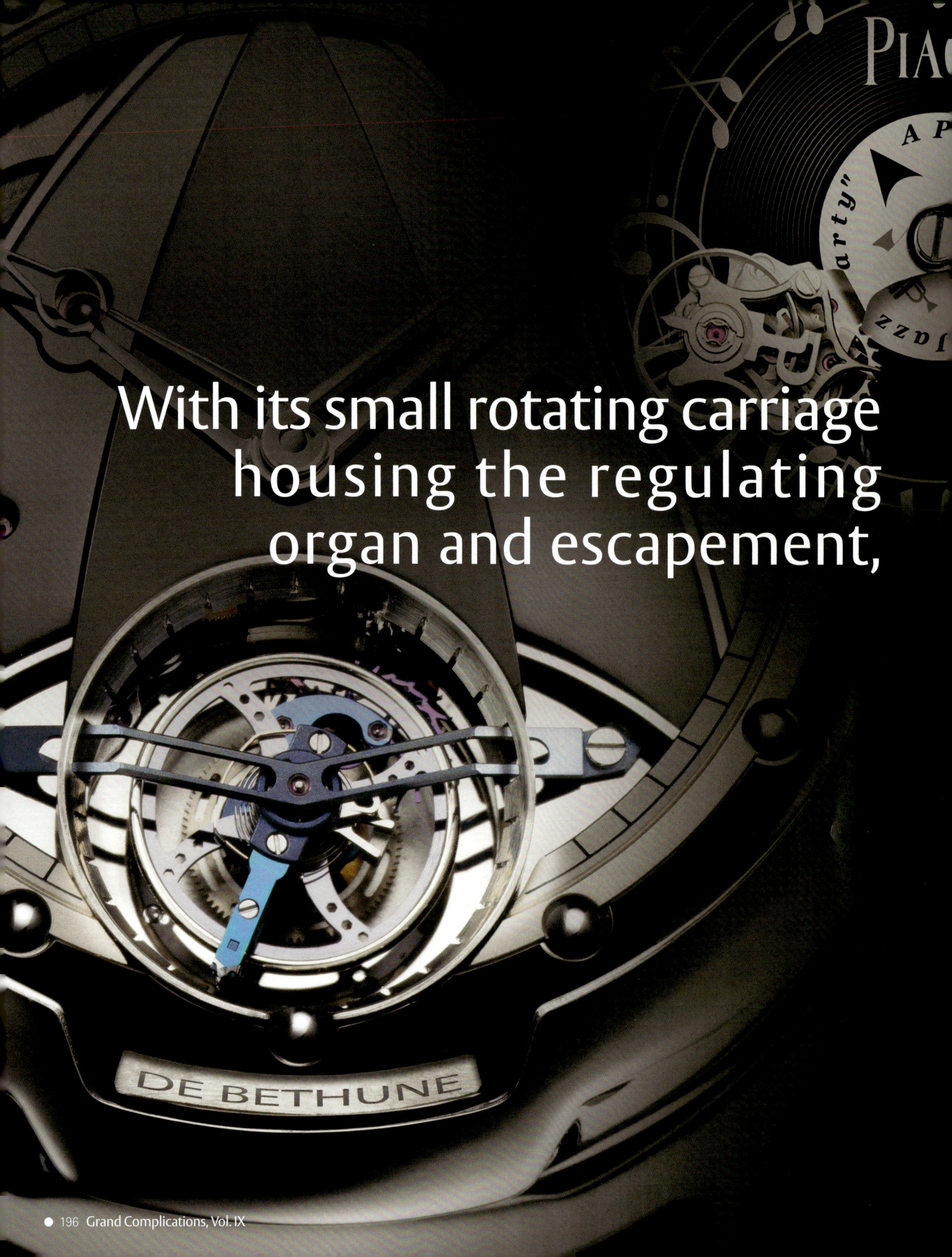

With its small rotating carriage housing the regulating organ and escapement,

the famous tourbillon is not just one of the most sophisticated horological mechanisms. It is also the most visually spectacular, the most magical, the most "photogenic" complication—all of which leads to an enduring fascination. Long limited to a very restricted production, the tourbillon has become a fashionable phenomenon—at the risk of becoming banal. However, today's watchmakers have undertaken to reinvent the mechanism, with architectures that play on the effects of transparency or extend into the third dimension, as well as ultracomplex systems that combine several rotational axes or tourbillons. Another marked trend is the rising popularity of tourbillon models designed for women's watches.

Pocket watches were usually carried in a special pouch or fob on a trouser belt, waistcoat or jacket—hence the name. Thus maintained in a vertical position for many hours, the mechanism—starting with the balance and the balance-spring, two necessary components for accuracy and precision—underwent the detrimental effects of gravity. To avoid such irregularities, the famous watchmaker Abraham-Louis Breguet (1747-1823)—one of the greatest horological geniuses of all time—had the idea of enclosing the "heart" of the watch (balance, balance-spring and escapement) within a small mobile carriage spinning on its axis within a predetermined time (generally one, four or six minutes). It was for this rotation that he chose to name his invention "tourbillon"—a term from astronomy synonymous with "planetary system" among 18th-century scientists. Breguet submitted the patent for his "watch compensating for all the inequalities that occur in the balance and balance-spring" on the 7th of Messidor in the 9th year of the Republican calendar—otherwise known as June 6, 1801. According to the house's archives, 35 tourbillon watches were sold between 1805 and 1823, and throughout the 19th century the tourbillon remained an extremely rare complication, reserved for a few exceptional timepieces.

The 20th-century triumph of the wristwatch could have brought about the disappearance of the tourbillon, as the complication loses much of its interest when the watch moves in all directions with the movement of the wearer's wrist. However, the revival of mechanical horology in the 1980s propelled this mechanism into the firmament of horological complications by positing it as the nec plus ultra of Swiss savoir-faire, conferring upon any watchmaker entry into the ranks of the great. Since the turn of the millennium, the tourbillon has become wildly popular. Whereas only 670 tourbillons were produced between 1801 and 1986, today the global production stands at over 1,000 per year. Over 50 Swiss brands offer them among their collections, which is surprising when one considers that only a small handful of artisans are really capable of making this small magic carriage. Faced with this (relative) banalization, the major watch firms are determined to offer "in-house" tourbillons featuring new technical and aesthetic developments.

TECHNIQUE

HOW DOES IT WORK?

Placed horizontally, a watch's regulating organ (balance, balance-spring and escapement) is uniformly affected by the Earth's gravitational pull. However, when it is placed vertically (and to a lesser extent, in any position besides the horizontal), gravitational force causes certain fluctuations in its functioning and oscillations. These small accelerations and decelerations cause irregularities in the watch's functioning. The idea behind the tourbillon is to place the regulating organ in a small mobile carriage that completes one rotation in a given time, usually one minute. By successively adopting all possible positions, the carriage allows the small irregularities to cancel each other out. The tourbillon carriage is a tiny construction that may contain 70 components with a weight of less than 1g. This is why the production, assembly and setting of a tourbillon watch require exceptional expertise, reserved for the most elite watchmakers and incompatible with mass production.

TOURBILLONS AND PRECISION

The tourbillon was created for the pocket watch, which spends most of its time in a vertical position. It is less necessary on a wristwatch, when the daily movements of the wearer's wrist play the same role as the tourbillon in putting the regulating organ in a variety of positions. Some people go so far as to say that the tourbillon is completely useless in its classic form. To solve this problem, a new generation of tourbillons has arrived—with diverse variations on the manners, axes and speeds of rotation—that supposedly offers enhanced performances for wristwatches. Nonetheless, even though the tourbillon is not in itself synonymous with increased accuracy, the expertise necessary to create such a watch, and the extreme care involved in its production, are both guarantees of a watch that offers superior performance.

FLYING TOURBILLON

Normal tourbillons are supported by two bridges, one above the mechanism and the other below. "Flying" tourbillons are supported by only one, below the mechanism, which allows the watchmaker to give the construction a more ethereal and resolutely virtuoso appearance.

TOURBILLON AND KARUSSEL

Patented by the Danish Bonniksen in 1892, the karussel has the same ultimate goal as the tourbillon. However, it is distinguished (or could be distinguished—the experts do not all agree) by the double wheel train that links it to the barrel: the first provides the necessary energy for the functioning of the escapement, and the second controls the speed of the carriage's rotation.

A Reinvented Classic

With its mobile carriage spinning on its axis like a planet, the tourbillon provides a fascinating sight that most brands are keen to exhibit through the dial, traditionally through an aperture at 6 o'clock. But the ultra-sophisticated mechanism is now the object of numerous technical and aesthetic innovations, such as the use of new materials that offer enhanced lightness, or the search for unusual dial configurations.

Jaquet Droz
Grande Seconde Tourbillon

Jaquet Droz has equipped its Grande Seconde Tourbillon with a sapphire carriage placed in the center of the seconds display. This watch, featuring an original dial configuration, is offered with a dial in Côtes de Genève or Grand Feu enamel.

Vacheron Constantin
Patrimony Traditionnelle 14-Day Tourbillon

Using four barrels in paired series, the Patrimony Traditionnelle 14-Day Tourbillon displays a power reserve of 336 hours. The tourbillon's carriage is adorned with the Maltese Cross, Vacheron Constantin's emblem.

Jaeger-LeCoultre
Master Ultra Thin Tourbillon

A model of restraint, Jaeger-LeCoultre's Master Ultra Thin Tourbillon (with an ultra-thin rose-gold case) gives the star role upon its eggshell-white dial to a large tourbillon within an ultralight carriage in grade-5 titanium.

Bulgari
Daniel Roth 8-Day Tourbillon

Besides its highly original design, the Bulgari Daniel Roth 8-Day Tourbillon stands out for its eight-day power reserve, displayed (as is the date) on the back of the case.

Breguet, *Marine Tourbillon Chronograph 5837*

A worthy heir to the famous invention by Abraham-Louis Breguet, the Breguet Marine Tourbillon Chronograph associates a balance-spring, escapement wheel and lever in silicon with a tourbillon featuring a carriage partially made of titanium.

Audemars Piguet, *Extra-Thin Royal Oak Tourbillon*

A tribute to the 40th anniversary of the Royal Oak, Audemars Piguet's Extra-Thin Royal Oak Tourbillon associates a movement just 4.46mm thick—one of the thinnest of its kind—with a "Petite Tapisserie" dial inspired by the first historic models.

Antoine Preziuso, *GTS Grand Tourbillon Sport*

Equipped with a cobalt chrome case, whose aerodynamic forms evoke the sportscars that inspired them, Antoine Preziuso's GTS Grand Tourbillon Sport is also notable for its large red hands, which recall the needles on speedometers.

Montblanc, *ExoTourbillon Chronographe*

On its ExoTourbillon Chronographe, from the Villeret 1858 collection, Montblanc has placed the balance on the exterior of the tourbillon carriage. As the balance is larger, its amplitudes gain more precision, and the tourbillon carriage, freed from the weight of the balance, requires less energy for its rotation.

Patek Philippe, *10-Day Tourbillon, Ref. 5101*

Endowed with a power reserve of ten days, the 10-Day Tourbillon, Ref. 5101 from Patek Philippe is notable for its cambered Art Deco-style case. As with all its tourbillon watches, the manufacture prefers not to show the tourbillon from the dial side, for fear that sunlight might dry up the lubricating oils.

Richard Mille, *Tourbillon RM 053 Pablo Mac Donough*

Designed to withstand the extreme shocks encountered during a polo match, Richard Mille's Tourbillon RM 053 Pablo Mac Donough (dedicated to one of the sport's most eminent stars) is equipped with two "windows," the left one for the tourbillon and seconds display, the right one for the hours and minutes.

Room for Transparency

Certain brands have chosen to distill the tourbillon's magic with constructions whose heart is open, revealing the entire interior of the movement—masterpieces of skeletonization that push the frontiers of possibilities, or fascinating transparent stage-settings within which the mobile carriage seems to float weightlessly.

Corum, *Golden Bridge Tourbillon*

Corum shows off its famous "baguette" movement in this Golden Bridge Tourbillon model, with a sapphire bridge. In order to reveal the flying tourbillon, a part of the movement (usually in gold) has been replaced by a sapphire component.

Franck Muller

Giga Tourbillon, version 8889 T G SQT BR NR

On Franck Muller's Giga Tourbillon, a large tourbillon of 20mm in diameter occupies the entire lower half of the dial. The two-barrel movement is equipped with a nine-day power reserve.

Piaget
Emperador Coussin Tourbillon Automatic Ultra-Thin

Equipped with an exceptionally thin manufacture movement, the Emperador Coussin Tourbillon Automatic Ultra-Thin from Piaget stands out for its unusual geometry, with an off-center hours and minutes display, tourbillon carriage at 1:30 and a micro-rotor visible on the dial at 9 o'clock.

Montblanc, *Tourbillon Bi-Cylindrique*

The Tourbillon Bi-Cylindrique from Montblanc stands out for its architecture with openworked heart, double cylindrical balance-spring and its "mysterious hours" display that seems to float in the air thanks to the use of crystal discs.

Haldimann, *H8 Sculptura*

On its H8 Sculptura, Haldimann has eliminated the hour and minute indications, retaining only the fascinating spectacle of the central tourbillon floating in between two sapphire crystals. It is a true work of art, constructed entirely with traditional techniques.

Revelation, *R01 Double Complication*

The first model from the brand Revelation, the R01 Double Complication mysteriously reveals its movement using a patented system of polarized lenses. It is also equipped with a Tourbillon Manège® movement, whose regulating organ is mounted on a central rotating mobile arm.

Girard-Perregaux, *Laureato Tourbillon with three bridges*

In 1889, Girard-Perregaux made watchmaking history by creating its famous "tourbillon with three gold bridges." In its Laureato Tourbillon with three bridges, the manufacture revisits this legacy by associating a titanium case with three translucent bridges in blue spinel.

Audemars Piguet
Jules Audemars Large Date Tourbillon

The Jules Audemars Large Date Tourbillon from Audemars Piguet, powered by a manual-winding in-house caliber, opts for understated elegance with a large tourbillon carriage and a panoramic date display at 12 o'clock.

Blancpain, *L-Evolution One-Minute Flying Sapphire Carrousel*

Blancpain has achieved a world first by reviving the carrousel (also known as the karussel), a mechanism that is close to the tourbillon (see "How Does It Work?"), and by incorporating it in several wristwatches, including the L-Evolution One-Minute Flying Sapphire Carrousel, with a tantalum case.

Harry Winston
Ocean Tourbillon Big Date

On the Ocean Tourbillon Big Date from Harry Winston, the tourbillon carriage and regulating organ that it houses seem to float freely in the air, with no mechanical support.

De Bethune, *DB 28T*

The DB 28T from De Bethune stands out for its futuristic architecture and its ultralight tourbillon in titanium and silicon. The tourbillon carriage rotates over 30 seconds and the balance oscillates at 36,000 vph.

The Third Dimension

Recent years have also witnessed the appearance of a new generation of tourbillons equipped with inclined carriages or several simultaneous rotational axes, and/or revolving at speeds that diverge from the traditional one minute (often 30 or 24 seconds). Besides their often spectacular appearance, these three-dimensional constructions are designed to be more suited to wristwatches, by multiplying the positions of the balance-spring with the case.

Harry Winston, *Histoire de Tourbillon 2*

The model Histoire de Tourbillon 2 from Harry Winston is equipped with a bi-axial flying tourbillon, with an outer carriage in titanium rotating over 120 seconds and an inner carriage inclined at a 10° angle turning over 40 seconds. The various indications—tourbillon paired with seconds at 9 o'clock, hours and day/night indication at 2 o'clock, minutes at 6 o'clock—are separated in off-centered subdials.

Concord, *Quantum C1 Gravity*

Concord created shock waves with its launch of the Quantum C1 Gravity, equipped with an unprecedented bi-axial tourbillon positioned vertically, i.e. almost outside the movement, and linked to the bottom plate by a series of cables reminiscent of suspension bridges.

Jaeger-LeCoultre, *Duomètre Sphérotourbillon*

Equipped with a Dual-Wing movement, Jaeger-LeCoultre's Duomètre Sphérotourbillon can rightfully claim to be the first tourbillon watch adjustable to the second. The "sphérotourbillon" rotates on two axes: one rotation is completed in 30 seconds, and the other (on a 20° angle) in 15 seconds.

Greubel Forsey
Tourbillon 24 Secondes Contemporain

On the Tourbillon 24 Secondes Contemporain from Greubel Forsey, the carriage—set at 25° and turning 2.5 times faster than an ordinary carriage—called for a new inclined gearing profile, as well as specific alloys.

Panerai
Radiomir Tourbillon GMT Ceramica Lo Scienzato

On the Radiomir Tourbillon GMT Ceramica Lo Scienzato from Panerai, the carriage turns at a right angle to the oscillation axis of the balance, performing two turns per minute. Its in-house Caliber P.2005/S is also distinguished by a substantial six-day power reserve, achieved thanks to the energy from three barrels.

Thomas Prescher
Triple Axis Tourbillon Regulator

On its Triple Axis Tourbillon Regulator, Thomas Prescher proposes a triple-axis flying tourbillon, supported on just one side—a real horological feat. The carriage rotating over one hour serves to indicate the minutes, while the hours and seconds can be read in two subdials.

Girard-Perregaux, *DLC Titanium Bi-axial Tourbillon*

The DLC Titanium Bi-axial Tourbillon by Girard-Perregaux combines two concentric carriages, spinning on their respective axes in 45 seconds and one minute and fifteen seconds respectively. The two carriages weigh only 0.8g, and total over 110 components.

Greubel Forsey, *Double Tourbillon 30°*

The esteemed tourbillon specialists Greubel Forsey have developed several innovative constructions that play on different angles and speeds of rotation—such as the Double Tourbillon 30°, whose inner carriage rotates in one minute and outer carriage rotates in four minutes.

Double Revolutions

A few innovative models offer a sophisticated system in which the tourbillon completes a rotation in one minute, while performing a complete turn of the dial in one hour—mirroring the double rotation of the Earth on its axis and around the sun. These are truly magical merry-go-rounds.

Piaget, *Polo Tourbillon Relatif Limelight Jazz*

The principle of a double revolution applies to Piaget's Polo Tourbillon Relatif Limelight Jazz. Suspended from the tip of the minute hand, the mobile flying tourbillon carriage appears to be disconnected from the movement driving it, creating a fine encounter between mechanical prowess and visual magic.

Cartier, *Calibre Astrotourbillon*

The Calibre Astrotourbillon from Cartier presents an innovative construction with a central tourbillon placed above the movement. The elongated carriage spins around the dial in one minute, thus enabling the arrow-shaped balance to indicate the seconds.

Ulysse Nardin, *Freak Diavolo*

The Freak Diavolo from Ulysse Nardin stands out for its very innovative construction. The movement rotating within the case allows the watch to display the hours and minutes with neither hands nor dial. The flying one-minute tourbillon indicates the seconds. The escapement and balance spring are in silicon.

Watches with Several Tourbillons

Another technical and aesthetic path explored by some watchmakers is the multiplication of tourbillons within a single watch. Horologers come up with double, even quadruple tourbillons, tied together by differential gear systems that average out the functioning of different regulating organs.

Breguet
Twin Rotating Tourbillons

The Twin Rotating Tourbillons from Breguet boasts two independent tourbillons coupled by a differential gear and fixed to a bottom plate completing a full rotation in 12 hours. These two mechanisms mutually compensate for their infinitely small variations in rate in order to offer superior precision.

Roger Dubuis
Excalibur Double Tourbillon Squelette

Boasting two flying tourbillons linked by a differential, the Excalibur Double Tourbillon Squelette model reveals all the technical sophistication and aesthetic refinement of its 292-component movement, graced by the prestigious Poinçon de Genéve.

Greubel Forsey, *Invention Piece 2*

A reinterpretation of its 2008 Quadruple tourbillon à différentiel, Greubel Forsey's Invention Piece 2 possesses two double tourbillon carriages (a first carriage executing a complete rotation in one minute, contained within a second carriage effecting a complete revolution in four minutes), linked by a spherical differential placed at the center of the timepiece.

Harry Winston, *Histoire du Tourbillon 3*

Released in a limited edition of 20 pieces, Harry Winston's Histoire du Tourbillon 3 watch presents three independent dial openings, one of which—on a slant—houses a spectacular bi-axial tourbillon equipped with two carriages, as well as a second tourbillon on a single axis.

TAG Heuer, *Carrera MikrotourbillonS*

The Carrera MikrotourbillonS from TAG Heuer possesses two tourbillons, one for the time, the other for the 1/100-of-a-second chronograph. The latter, described as "the fastest tourbillon in the world," beats at 360,000 vph and completes 12 revolutions per minute.

Jacob & Co.

Napoleon Quadra Tourbillon

Crafted in titanium, the imposing Napoleon Quadra Tourbillon from Jacob & Co. features no fewer than four skeletonized tourbillons and four complete time zones with hour and minute displays.

The Feminine Tourbillon

Over the last few years, women's tourbillon watches have accomplished an astonishing breakthrough. Uniting haute horology and—most often—haute jewelry, the most prominent brands have undertaken to create timepieces that set a refined stage for the little "magic carriage."

Jaeger-LeCoultre, *Rendez-Vous Tourbillon Enamel*

The Rendez-Vous Tourbillon Enamel from Jaeger-LeCoultre gives pride of place to rare artisanal skills, with an enameled dial depicting a rose and its double heart. It is powered by an automatic-winding manufacture movement with tourbillon.

de Grisogono, *Tondo Tourbillon Gioiello*

Adorned with more than 500 diamonds, the Tondo Tourbillon Gioiello from de Grisogono houses within its white-gold oval case an exclusive movement, with off-centered tourbillon and 8 o'clock. The strap is in galuchat, and the face is covered with mother-of-pearl.

Blancpain, *Tourbillon Octopus*

Set with rubies arranged to resemble the tentacles of an octopus (hence the name), the Tourbillon Octopus from Blancpain's women's collection also displays a large date in a double aperture at 6 o'clock.

Audemars Piguet
Jules Audemars Ladies' Tourbillon Chronograph

The Jules Audemars Ladies' Tourbillon Chronograph from Audemars Piguet reveals its tourbillon on a mother-of-pearl dial that is set with pavé diamonds. The white-gold case sparkled with 251 brilliant-cut diamonds and 68 baguette-cut diamonds.

Parmigiani, *Kalpa XL Tourbillon Diamants*

The Kalpa XL Tourbillon Diamants offers a one-week power reserve and a carriage performing two rotations per minute. The tonneau-shaped case, inspired by the golden ratio, is adorned with a precious diamond pavé setting.

Van Cleef & Arpels, *Midnight Tourbillon Nacre*

The superbly understated Midnight Tourbillon Nacre from Van Cleef & Arpels is available in a limited edition of 100 watches. The handcrafted mother-of-pearl dial reveals a tourbillon carriage that echoes the style of the Vendôme Column in Paris.

Cartier, *Tourbillon & Crocodile Watch*

On the haute jewelry Tourbillon & Crocodile Watch, Cartier has accompanied one of the stars of its bestiary with a flying tourbillon whose C-shaped carriage indicates the seconds. This work of art is highlighted by 585 gems.

Richard Mille

RM 051 Phoenix – Michelle Yeoh

The RM 051 Phoenix – Michelle Yeoh pays tribute to the mythical bird, with an interlacing diamond design linking the barrel, power reserve indicator and tourbillon. The watch was designed by Richard Mille in collaboration with its namesake star.

A. LANGE & SÖHNE

LANGE 1 TOURBILLON PERPETUAL CALENDAR – REF. 720.025

The Lange 1 Tourbillon Perpetual Calendar's self-winding L082.1 caliber is ingeniously fitted with a platinum centrifugal mass for exceptional winding efficiency. Thanks to an innovative rotating peripheral month ring that occupies a minimal amount of space, and an hour and minute circle off-centered on the solid silver rhodiumed dial, A. Lange & Söhne accomplishes a perpetual calendar of remarkable ease of use and legibility, despite its comprehensive constitution of moonphase, retrograde day, leap-year aperture and date. Admired intimately through the sapphire crystal caseback of this 41.9mm platinum masterpiece, the exquisite tourbillon is embellished with an extremely sophisticated black polished finish, and boasts a flawless diamond end stone as its radiant bearing.

A. LANGE & SÖHNE

RICHARD LANGE TOURBILLON POUR LE MERITE – REF. 760.032

The Richard Lange Tourbillon "Pour le Mérite" is powered by the manual-winding new hand-finished Lange manufacture L072.1 caliber. It incorporates a fusee-and-chain transmission as well as a tourbillon with a patented stop-seconds mechanism and a balance spring manufactured in-house. The 41.9mm case comes in platinum or rose gold with a 36-hour power reserve. Its dial features three intersecting circles for the time indicators. The largest dial indicates the minutes while the smaller subdials represent the seconds and hours. The watch includes a crocodile strap secured with a solid gold buckle. The Richard Lange Tourbillon "Pour le Mérite" in platinum is limited to 100 pieces.

AUDEMARS PIGUET

JULES AUDEMARS DRAGON TOURBILLON – REF. 26569OR.OO.D088CR.02

The Jules Audemars Dragon Tourbillon is powered by Audemars Piguet's in-house, hand-wound Manufacture caliber 2906, which is comprised of 215 parts and beats at 21,600 vph. This timepiece has an 18K pink-gold case and pink-gold enamel dial that features an engraved 18K pink-gold dragon. The strap is hand-stitched, large square scale crocodile with an 18K gold AP folding clasp. It is water resistant to 20m. There is a minimum guaranteed power reserve of 72 hours. This is a limited edition of 18 pieces.

AUDEMARS PIGUET

JULES AUDEMARS LARGE DATE TOURBILLON – REF. 26559OR.OO.D002CR.01

Home to Audemars Piguet's new 228-part 2909 caliber, this masterpiece showcases, via a stunning aperture at 6:00, a tourbillon that unarguably merits its prominent exhibition on the dial. The exposed heartbeat of the watch is balanced by the presence of an innovative large date at 12:00. The ingenious use of two superimposed dials, working in precise synchronization and without a separation of the tens and units, is crucial in permitting a date display of such magnitude. The Jules Audemars Large Date Tourbillon boasts a 72-hour power reserve, beats at a frequency of 21,600 vph, and offers the choice of black or silver-toned dial, each of which is limited to 25 pieces.

AUDEMARS PIGUET

ROYAL OAK EXTRA-THIN TOURBILLON – REF. 26510OR.OO.1220OR.01

The Royal Oak Extra-Thin Tourbillon is powered by the hand-wound, in-house Manufacture caliber 2924, enclosed in an 18K pink-gold 41mm case. The movement is comprised of 216 parts and beats at 21,600 vph. It features a blue dial with the "Petite Tapisserie" pattern and pink-gold applied hour markers. This piece includes the following functions: tourbillon, hours, minutes and power reserve. The bracelet is also made out of 18K pink gold with an AP folding clasp. It is water resistant to 50m and has a minimum power reserve of approximately 70 hours.

AUDEMARS PIGUET

ROYAL OAK OFFSHORE HAND-WOUND TOURBILLON WITH CHRONOGRAPH REF. 26288OF.OO.D002CR.01

The Royal Oak Offshore tourbillon chronograph, developed and crafted entirely by Audemars Piguet, houses the manual-winding 2912 caliber with two barrels, a tourbillon and a 10-day power reserve. The case is in 18K pink gold with a forged carbon bezel, black ceramic crown and pushers, assembly screws in polished steel and a glareproof sapphire crystal and caseback. The black dial is snailed and openworked at 6:00, 9:00 and 12:00, with applied luminescent gold numerals and hands.

BELL & ROSS

BR 01 TOURBILLON AIRBORNE

This limited edition of 20 pieces has a manual-winding mechanical movement and contains carbon-fiber bridges and plates and a black-gold tourbillon carriage. Hours can be read in a counter at 12:00, a trust index is positioned at 3:00, and the 120-hour power reserve indicator is at 9:00. A central hand displays the minutes. The three-dimensional dial has a black mesh-like design and is treated with a photoluminescent coating for nighttime reading. The 46x46mm case is titanium with a black DLC (Diamond-Like Carbon) coating. Water resistant to 100m, the watch features a shagreen strap.

BELL & ROSS

BR MINUTEUR TOURBILLON PINK GOLD

This tourbillon is powered by a manual-winding mechanical movement and its bridges are in pink gold and anodized aluminum. The watch displays the hours, minutes, small seconds, a three-day power reserve indicator and a flyback function (2 counters: 60 minutes and 10 tenths of an hour), which features a zero-setting and quick reset. The case is in 18K pink gold and measures 44x50mm. The black dial possesses hands and indexes treated with a photoluminescent coating for nighttime reading. Water resistant to 50m, the watch is available on a rubber or alligator strap. Limited edition of 30 pieces.

BELL & ROSS

BR MINUTEUR TOURBILLON TITANIUM

This tourbillon is powered by a manual-winding mechanical movement and its bridges contain carbon-fiber mainplates and a black-gold tourbillon carriage. The watch displays the hours, minutes, small seconds, a three-day power reserve indicator and a flyback function (2 counters: 60 minutes and 10 tenths of an hour), which features a zero-setting and quick reset. The case is titanium with DLC (Diamond-Like Carbon) coating and measures 44x50mm. The black dial possesses hands and indexes treated with a photoluminescent coating for nighttime reading. Water resistant to 50m, the watch is available on a rubber or alligator strap. Limited edition of 30 pieces.

BLANCPAIN

L-EVOLUTION TOURBILLON LARGE DATE – REF. 8822-15B30-53B

Available in satin-brushed white- or red-gold versions, the L-evolution Tourbillon Large Date features the characteristic dial of this collection, in a color matching that of the case. In most watches, the power reserve indication is displayed on the watch dial, or on the bridge side of the movement; however, Blancpain offers a third solution, indicating the power reserve on a subdial placed directly on the oscillating weight. With this solution, the oscillating weight plays a role beyond that of merely transmitting energy: it becomes an information medium. This original device is incorporated within the Manufacture Blancpain 4225G caliber endowed with a 7-day power reserve. The movement, visible through the sapphire crystal caseback, also features Blancpain's "flying" tourbillon and double-disc large date mechanism.

BREGUET

MARINE TOURBILLON CHRONOGRAPH – REF. 5837PT.U2.5ZU

Housed within a 42mm case in 950 platinum, the 554.4 caliber, with silicon lever and escape wheel, expresses the magnificence of its regulatory organ through a generous aperture in the hand-engraved silvered, platinum-coated gold dial. Interrupting the face's distinguished wave motif, the timepiece's tourbillon, whose lightweight titanium carriage permits an exceptional resistance to shock and magnetism, astonishes beneath the watch's central blued steel Breguet hands and chronograph stem. Adjusted in six positions, and perfected with a Breguet balance wheel, this hand-wound masterpiece is finished with 30-minute and 12-hour totalizers at 3:00 and 6:00, as well as wave-shaped chronograph pushbuttons.

BREGUET

TRADITION FUSEE TOURBILLON WITH SILICON BALANCE SPRING – REF. 7047PT.11.9ZU

An apogee of Breguet's relentless innovative spirit and astonishing technical finesse, this hand-wound 41mm timepiece in 950 platinum echoes its extraordinary mechanical sophistication with a fascinating and expressive design. Boasting optimal torque regularity thanks to the connection of a fusée-and-chain transmission to the barrel, the 43-jewel 569 caliber with silicon Breguet balance spring reveals its 60-second tourbillon in grand fashion at 1:00. The recipient of one of the timepiece's numerous patents, the power reserve indicator is directly mounted on the exposed barrel at 9:00. Aesthetically, the anthracite-toned finish of the movement was ingeniously achieved via a sophisticated electrodeposition process that further speaks to the manufacture's endless pursuit of watchmaking perfection.

CHANEL

J12 TOURBILLON

After transforming ceramic into a precious material, CHANEL goes even further and brings ceramic to the very heart of its movement. The J12 Tourbillon is the first timepiece to feature a bottom plate made of high-tech ceramic, with 1/100 accuracy never before achieved in this material. It is the perfect blend of J12 high technology and one of the most respected watchmaking complications. Limited to 12 pieces, the J12 Tourbillon is equipped with the CHANEL 05-T2 (white version) or the CHANEL 05-T1 (black version) manual-winding movements, which provide a 100-hour power reserve. The J12 Tourbillon is available in 18K white gold and black high-tech ceramic or in 18K white gold and white high-tech ceramic, set with 46 baguette-cut diamonds (approx. 4.7 carats).

CHANEL

J12 TOURBILLON CHANEL-05-T1 – TECHNICAL SPECIFICATIONS

MOVEMENT
-Caliber: 10 1/2 lines
-Manual winding
-Bottom plate in high-tech* ceramic
-Thickness: 5.3mm
-Casing dimension: 28.5mm
-Number of jewels: 17
-Vibrations per hour: 21,600

MOVEMENT BRIDGES
- Hand drawn, bevelled and satin-finished by hand
- Black chromium finish for the movement CHANEL 05-T1
- Rhodium finish for the movement CHANEL-05-T2

* Haute technologie / High technology

OSCILLATING ORGAN
-Balance diameter: 9.7mm

MOTOR ORGAN
-Mainspring torque: 350g.mm
-Power reserve: 100 hours

TOURBILLON CARRIAGE
-Bridge in 18K white gold
-Upper mobile frame in optical organic glass
-Carriage weight: 0.47g

CHANEL

PREMIERE FLYING TOURBILLON

CHANEL makes its first foray into the world of feminine mechanical complication watches with its Première Flying Tourbillon. The flying tourbillon discreetly beats within its stylized camellia-shaped case, a tribute to Mademoiselle Chanel's favorite flower. The absence of an upper bridge magnifies the tourbillon's aesthetic—its frame becomes a micro-mechanical marvel in a three-dimensional assemblage of thin elements. This exceptional tourbillon was developed in collaboration with Renaud & Papi Manufacturing (APRP SA), the high-end research and development wing of Audemars Piguet. The 18K white-gold and ruby version is a one-of-a-kind piece, set with 181 baguette-cut rubies (approx. 12.74 carats), 1 rose-cut ruby (approx. 0.45 carat) and 34 brilliant-cut diamonds (approx. 0.034 carat). The 18K white-gold and diamond version is available in a limited edition of 20 pieces, and set with 101 baguette-cut diamonds (approx. 5.7 carats) and 127 brilliant-cut diamonds (approx. 2 carats).

CHANEL

PREMIERE FLYING TOURBILLON

- Caliber: "Camélia Flying Tourbillon"
- Movement designed exclusively for Chanel by Renaud & Papi (APRP SA)
- Mechanical manual-winding movement
- Power reserve: 40 hours
- Number of parts: 225
- Parts chamfered, drawn and circular-grained by hand
- Number of jewels: 18
- Flying tourbillon: 73 parts – frame mass 0.432g
- Frequency: 21,600 vph (3Hz)
- Motor unit: 1 fixed clamp barrel
- Mainspring torque: 720g.mm
- Custom-tailored rectangular bottom plate
- Anti-shock system on the balance
- Variable-inertia balance

DE BETHUNE

DB25 T

While this timepiece's star-studded blued titanium dial—with gold stars, central jumping seconds, and movement-operating indicator at 12:00—indicates a degree of elegance and understated beauty, the watch's caseback reveals a secret of sophistication and technical innovation. Visible only on the back of the 44mm white-gold wristwatch, the 36,000 vph DB 2119 caliber's tourbillon is a veritable triumph of horological mastery. Weighing a mere 0.18g, and completing its axial rotation once every thirty seconds, the silicon/titanium tourbillon, the lightest on the market, achieves an exceptional regulatory compensation of the laws of physics and the wrist's often erratic movements—a breakthrough that pays homage to Abraham-Louis Breguet's trailblazing invention.

DE BETHUNE

DB28 ST

Revealing the hand-wound DB 2119 caliber's hand-decorated mainplate with Côtes de Bethune motif, this mirror-polished titanium timepiece's dial sets the stage for a groundbreaking tourbillon, joining the blued steel indications of hours, minutes and jumping seconds. Surrounded by a frame indicating its 30-second rotation cycle, the 64-component silicon/titanium tourbillon, showcased at 6:00, beats at a frequency of 36,000 vph and weighs a record-setting 0.18g. Boasting a self-regulating twin barrel, as well as silicon/white-gold balance wheel, this 43mm timepiece displays an astounding level of sophistication within an architecture of befitting intricacy and individuality.

DE BETHUNE

DB28T

Powered by the 298-component DB 2019 caliber oscillating at a frequency of 36,000 vph, hand-decorated with circular-grained parts and Côtes de Bethune, this expressive timepiece boasts a tourbillon of exceptional distinction. Prominently positioned at 6:00, the silicon/titanium regulator performs its complete revolution in precisely 30 seconds. Weighing four times less than a conventional carriage, the timepiece's tourbillon excels with its minimal mass and inertia, as well as high frequency and rotation velocity. Embellished with a black polished stainless steel dial, the 42.6mm watch is worn on an extra-supple alligator leather strap.

de GRISOGONO

TONDO TOURBILLON GIOIELLO

An exquisite balance of mechanical excellence and luxury, this feminine masterpiece wears a sea of diamonds as it adorns its spectacular tourbillon at 8:00. The generously sized white-gold case, fully pavé with white diamonds, leads the eye to the black diamond set in the crown. Upon the mother-of-pearl dial, the off-center tourbillon joins two white-gold dauphine hands, powered by the DG 31-88 caliber. Worn on a white galuchat strap with diamond-set butterfly buckle, the Tondo Tourbillon Gioiello, like the woman who wears it, finds equal radiance in visual beauty and internal excellence.

F.P. JOURNE

TOURBILLON SOUVERAIN

The new Tourbillon Souverain possesses the aesthetic characteristics and technical demands of all F.P. Journe creations. Its mechanism uses the remontoir or constant-force device system, and has now been enriched with an independent seconds system, which provides a more accurate time read-off, as the seconds hand does not move until the second has fully elapsed. This exceptional mechanism is housed in an equally noble 38 or 40mm case in 18K white or red gold, and the technical and aesthetic feat is revealed through the transparent sapphire crystal caseback. The face features the distinctive identity of F.P. Journe chronometers, with an 18K red- or white-gold dial and guilloché silver subdials screwed to the watch face (a patented feature). The 42-hour power reserve indication at 12:00 perfectly balances the independent seconds display at 6:00.

FRANCK MULLER

GIGA TOURBILLON MATRIX

This remarkable timepiece is perfectly named as it houses one of the largest tourbillons in the world. The Giga Tourbillon Matrix showcases the brilliance and audacity of Franck Muller's research and development team. Its Ø 20mm tourbillon, which takes up about one half of the watch, is powered by four Ø 16mm barrels. These extra-wide barrels are paired in parallel series that double the power of the operating reserve. This results in a continuous and reliable force. This movement has been reversed, with the bridges placed on the dial side, making the watch a rare gem. The incomparable tourbillon is housed in an equally unparalelled and stunning cage. The 18K red-gold case and cage are completely studded with over 10 carats of diamonds, making this watch perfect for a watch enthusiast with pizzazz and a stylish flair.

FRANCK MULLER

THUNDERBOLT TOURBILLON

The Thunderbolt Tourbillon, also known as the Tourbillon Rapide, is the world's quickest tourbillon mechanism. The tourbillon cage, which makes one full rotation every five seconds is powered by four barrels. In other words, the complication rotates at a whopping 12 times per minute! The rotating frame is distinguished by a patented FM escapement with an escapement wheel, balance wheel, reversed anchor and of course, Breguet hairspring. Entirely manufactured in house, this speedy model's movement beats at 21,600 vph with a power reserve of 60 hours. It is also engraved, rhodium-plated, grained and hand beveled. This intricate movement is visible through a skeletonized dial. Its strong, sleek case is the perfect complement to this horological masterpiece.

FREDERIQUE CONSTANT

SLIMLINE TOURBILLON MANUFACTURE – REF. FC-980V4SZ9

Frederique Constant proudly presents its new limited edition with this sophisticated, subtle and ultra-slim timepiece. Its namesake complication, positioned at 6:00, adds dynamism and depth to the silvery, rounded dial. The seconds hand plays a key role in the 60-second rotating tourbillon cage, which displays both captivating grace and impeccable engineering. The slender watch, which measures less than one centimeter in thickness, is available in stainless steel or stainless steel with an 18K rose-gold bezel. Within this sleek case beats a movement in silicium, whose advantages include durability, extreme accuracy, and simplified maintenance. Each version is available in a limited and numbered edition of 188 pieces.

FREDERIQUE CONSTANT

FREDERIQUE CONSTANT MANUFACTURE GRAND FEU – REF. FC-980EGF4H9

Individually numbered and perfected with a silicon escapement, this 188-piece limited edition masterpiece's innovative self-winding FC-980 caliber boasts a perfect weight equilibrium within the 80-component tourbillon cage, thanks to the house's patented Smart Weight Balancing system. Adding to the shine of the 42mm 18K rose-gold case, the dial, enameled with the genuine Grand Feu technique, achieves the depth of its luster via a rare and ancient process that demands the highest degree of expert craftsmanship. While the splendid one-minute tourbillon with running seconds is revealed on the lower half of the dial, a rose-gold and blued-steel day/night indicator, read off an Arabic numeral inner semicircle, ingeniously joins the black steel hour and minute hands. Completed with a sapphire crystal caseback and hand-stitched alligator leather strap, this exclusive timepiece exquisitely blends the brilliance of the past with the triumph of modern innovation.

GIRARD-PERREGAUX

DLC TITANIUM BI-AXIAL TOURBILLON – REF. 99810-24-000-BA6A

The astonishingly modern architecture of Girard-Perregaux's iconic movement is here enhanced by a black DLC-coated titanium case and magnified by the contrast of colors and materials. The Girard-Perregaux Bi-Axial Tourbillon combines two concentric cages that enable the regulator to make multi-dimensional rotations. This complex construction compensates for rate irregularities due to gravity. An internal cage bearing the balance wheel, balance spring and escapement completes one full turn on its axis in 45 seconds. In addition, an external cage completes one revolution in one minute and 15 seconds, enabling a rotation on its second axis. A full revolution is completed in three minutes and 45 seconds. The sophisticated mechanism has an astounding delicacy: the two cages weigh just 0.8g, yet comprise a total of over 110 components.

GIRARD-PERREGAUX

TOURBILLON WITH THREE GOLD BRIDGES – REF. 99193-52-001-BA6A

The expression of a unique heritage whose origins can be traced back to the mid-19th century, this new version in pink gold is available in a strictly limited edition of only 50 pieces. Contained within its diameter of 41mm, its three entirely hand-finished, arrow-shaped bridges attract all eyes. The tourbillon device is also the subject of particular attention, its 72 component parts being meticulously assembled within a space only 12mm in diameter. The upper and lower cages are beveled and hand-polished, as are the bridges. Under the sapphire crystal, the three bridges carved out of pink gold form a striking contrast with the color of the platinum. This combination of materials gives the timepiece, powered by the famous GP 9600 self-winding movement, a contemporary air of distinction and a strongly marked character.

GLASHÜTTE ORIGINAL

GRANDE COSMOPOLITE TOURBILLON – REF. 89-01-03-03-04

The exquisitely complicated Grande Cosmopolite Tourbillon enables the world traveler to track the time in any two of 37 world time zones. The 48mm platinum case frames a silver-grained, solid 18K gold dial, which displays destination time in the center and home time at 6:00. The unique combination of complications, which includes dual time zones and perpetual calendar, is crowned by the extraordinary Flying Minute Tourbillon. Glashütte Original has applied for four separate patents to protect the expertise that made this unique work of art possible. A sapphire crystal caseback offers a clear view of the manual-winding manufacture Calibre 89-01 with more than 500 components. The caseback is protected by a platinum hunter case fitted with a graphic presenting 37 time zones. This exclusive masterpiece is limited to 25 pieces worldwide.

GUY ELLIA

TOURBILLON MAGISTERE II

Guy Ellia has renewed its collaboration with Christophe Claret to present their latest timepiece enhanced with a tourbillon escapement: the Tourbillon Magistère II. Inside its distinctive case, an elongated rectangle (44.2x36.7mm), the skeleton movement features mysterious winding and features an "invisible" dial, whose main hands display the hours and minutes while seeming to float in space. In fact, the technical genius of this watch, produced in a limited edition of 12 pieces for each version, lies in the inversion of its bridges compared to a classic tourbillon timepiece. Hence the location of the winding crown at 6:00 and the use of the crown positioned at 12:00, while maintaining pressure on the security pushbutton at 10:00 to set the time. This piece is for lovers of fine watchmaking in search of the unusual.

GUY ELLIA
TOURBILLON MAGISTERE TITANIUM

Designer Guy Ellia presents the titanium version of his famous skeletonized mysterious-winding tourbillon. The Time Square Tourbillon Magistère's Caliber TGE 97 was created by Swiss manufacturer Christophe Claret and features a 110-hour power reserve, one-minute tourbillon, and a titanium bottom plate and bridges. The movement beats at 21,600 vph within an entirely hand-chamfered cage.

GUY ELLIA

TOURBILLON ZEPHYR

With a movement created by Swiss manufacture Christophe Claret, the Tourbillon Zephyr symbolizes transparency with excellence. Caliber GES 97 features a 110-hour power reserve and winding ring set with 36 baguette-cut diamonds (1.04 carats), or engine turning and one-minute tourbillon movement. The latter beats at 21,600 vph within an entirely hand-chamfered cage. The convex case with 950 platinum sides has a transparent back cover and hearth engraving. Mounted on an alligator strap with a folding buckle, this model is also available in platinum and titanium.

HARRY WINSTON

HISTOIRE DE TOURBILLON 3 – REF. 500/MMTWZL.K

Harry Winston's Histoire de Tourbillon collection celebrates the tourbillon as a scientific work of art. This particular model choreographs two rotating escapements that seem to defy gravitational laws. Three tourbillons, two of them nested inside one another, rotate at different speeds: one carriage rotates every 40 seconds while the second rotates every 120 seconds on an axis perpendicular to the first. The third tourbillon rotates separately on a more traditional single axis, yet it is the fastest of the three cages with one revolution per 36 seconds. To maximize precision, the two regulating systems are connected through a gear exemplifying an extremely delicate exercise in the balance of power. The case, composed of 77 parts in white gold, Zalium™ and non-reflective sapphire crystal, beautifully displays the unique mechanisms inside. This remarkable timepiece is available in a limited edition of 20 pieces.

HARRY WINSTON

OCEAN TOURBILLON BIG DATE – REF. 400MMTBD45WL.K

The Harry Winston Ocean Tourbillon Big Date is a masterpiece with an unrivaled, aesthetic take on the classic complication. Generally, the tourbillon is mounted at the plate of the watch and is only visible from one angle. This model radically differs from traditional tourbillon construction as Harry Winston dared to create a "floating tourbillon," which essentially means that the carriage appears to float freely with no apparent connection to the rest of the watch's movement. Harry Winston mastered this craft by using sapphire crystal on both sides of the model, as well as a carefully designed bridge that apparently suspends the cage in empty space.

HUBLOT

CLASSIC FUSION SKELETON TOURBILLON – REF. 505.OX.0180.LR

While this self-winding tourbillon timepiece pays homage to the brand's history, its modernity is without question. The 18K King Gold of the case, an alloy developed by Hublot by adding 5% platinum to the mass of melting gold, produces an intensity of color reinforced through contrast with the vibrant black of the watch's strap and internal components. Shaped in reference to Hublot's very first creations, the 18K red-gold-plated hour and minute hands revolve above the skeleton backdrop of the self-winding MHUB6010.H1.8 caliber and its gorgeous tourbillon at 6:00. This 99-piece limited edition is worn on a black alligator strap stitched onto black rubber.

HUBLOT

TUTTI FRUTTI TOURBILLON – REF. 345.PL.5190.LR.0901

This paragon of complications has been infused with incurable happiness. Bringing a tourbillon to the light-hearted feminine genius of the Tutti Frutti collection, Hublot lets the joy of color and diamonds enchant the vivacious brilliance of the treasured complication. With a bezel set with 48 baguette-cut blue sapphires on a 41mm 18K red-gold case set with 198 diamonds, the Tutti Frutti Tourbillon forms a luxurious and delightful home for the raised flying tourbillon—which acts without ball bearings—to reign. Combining the dazzling shimmer of 621 diamonds and the watchmaking excellence of its HUB6004 hand-wound movement, Hublot makes a modern declaration about the meaning of ultra-feminine.

JACOB & CO.

RAINBOW BLACK DIAMOND BAGUETTE TOURBILLON – REF. R13

The Rainbow Black Diamond Baguette Tourbillon possesses not only a manual-winding tourbillon, but a sophisticated world-time function split among three subdials, featuring a total of 25 time zones, each one indicated by the name of a reference city. Powered by the JCO Caliber 1 with a power reserve of 120 hours, this impressive piece is housed within an 18K white-gold case. The sleek dial is set with a total of 5.46 carats of invisibly set baguette-cut black diamonds, and 30.06 carats of black diamonds adorn the case, bezel and buckle upon an alligator strap.

JACOB & CO.

QUENTTIN – REF. QENMBK

The Quenttin Caliber 5 with vertical tourbillon boasts a power reserve of 744 hours—31 days—that would be exceptional in any mechanical timepiece, let alone a tourbillon. Seven barrels, juxtaposed vertically and mounted co-axially, hold the key to this extraordinary autonomy. The manual-winding movement within is made entirely in Switzerland and positioned vertically, just like the original tourbillon design by Abraham-Louis Breguet. The Quenttin's movement is revolutionary because the wheels do not revolve around a central arbor, but around several arbors aligned horizontally, like the gearbox in a car engine. This leads to a unique layout in which the time and power reserve indications are displayed via cylinders rather than hands.

JAEGER-LECOULTRE

DUOMETRE SPHEROTOURBILLON – REF. 605.25.20

A member of the highly exclusive Dual-Wing family, the Duomètre Sphérotourbillon displays its allegiance in several ways. The dial geometry reflects the nature of the Dual-Wing system, with the hours, minutes and date across from the tourbillon, two separate power reserve displays (one for each barrel), and two small subdials at the top and bottom of the dial, indicating the seconds and 24-hour scale, respectively. A pushbutton at 2:00 uses a flyback mechanism to send the seconds hand back to zero, making the Duomètre Sphérotourbillon the first tourbillon model that is adjustable to the second—a marvel of precision. The tourbillon itself revolves not only around the axis of its titanium cage, but also a second axis, inclined at 20°. This unusual rotation is also quite fast: 30 and 15 seconds, respectively. The transparent caseback reveals Jaeger-LeCoultre Calibre 382, a stunning hand-assembled and -decorated movement.

PANERAI

LO SCIENZATO – RADIOMIR TOURBILLON GMT CERAMICA – REF. PAM00350

Powered by the in-house P.2005/5 hand-wound caliber with three barrels providing a power reserve of six days, this 48mm timepiece achieves an exceptional hardness and corrosion resistance thanks to its sophisticated construction in black zirconium oxide ceramic. Visible through the expressive dial, the 277-component movement boasts a 30-second tourbillon at 10:30 that provides distinguished precision to an active mechanical demonstration as it rotates perpendicularly to the axis of the balance. Along with an am/pm indicator at 3:00, and small seconds/tourbillon movement counter at 9:00, this 31-jewel wristwatch boasts a second time zone indicator via a central arrow pointer as luminescent as the hour and minute hands.

PARMIGIANI

KALPA TOURBILLON 0 – REF. PFH150-1000200-HA1241

Tonneau-shaped and crafted in 18K rose gold, the Kalpa Tourbillon 0 adorns the mechanical transparency of its open-worked center with an outer stage decorated with Côtes de Genève. At 12:00, standing opposite the 30-second tourbillon at 6:00, a hand-driven scale indicates the status of the hand-wound PF 501 caliber's impressive seven-day power reserve. Limited to 30 pieces, as engraved on the caseback, the 44.7x37.2mm timepiece drives, from its revealing center, a combination of seconds, and hour and minute hands enhanced with luminescent coating.

PARMIGIANI

TORIC TOURBILLON 40 – REF. PFH436-1002800-HA1241

Joining the remarkably legible small seconds and 30-minute chronograph counters of mirrored aesthetic identity, this 10-piece, engraved, "Édition Spéciale" hand-wound timepiece's magnificent tourbillon regulator rises from the expressive dial layout to illustrate the watch's devotion to exceptional precision. Contrasting the tones and imposing stature of the 46mm 18K rose-gold case with double knurled bezel, the slender chronograph seconds stem is met by two refined blue javelin-shaped hands with luminescent coating. The Toric Tourbillon is worn on an Hermès alligator leather strap.

PATEK PHILIPPE

MEN'S GRAND COMPLICATIONS – REF. 5101J

While the slender yellow-gold rectangular case tells a story of elegant refinement, this timepiece's TO 28-20 REC 10J PS IRM manual-winding caliber counters with a statement of raw power and extraordinary mechanical finesse. Boasting a tourbillon with an astonishing 10 days of power reserve, indicated in a counter at 12:00, the 231-component movement is equipped with two mainspring barrels. On the understated, grained dial with applied gold Arabic numerals, the hours and minutes are complemented with a small seconds counter at 6:00. Finished with a revealing sapphire crystal caseback, the watch is worn on a hand-stitched, dark brown alligator strap.

PATEK PHILIPPE

MEN'S GRAND COMPLICATIONS – REF. 5207R

Beneath the timepiece's uncluttered black dial, the RTO 27 PS QI caliber uses the full complement of its 549 components to perform a choreography of phenomenal sophistication. Activated by a slide-piece on the left side of the 41mm rose-gold case, a two-gong minute repeater brings sonorous delight to the visual impact of the watch's expressive décor. Regulated by its tourbillon escapement, the hand-wound movement with flywheel provides, with optimal reliability and accuracy, a perpetual calendar with additonal day/night indicator at 7:30. Three apertures on the upper section of dial display the day, date and month of the instantaneous calendar, while the leap-year status is positioned to the right of the moonphase display and small seconds subdial at 6:00.

PIAGET

PIAGET EMPERADOR COUSSIN XL TOURBILLON – REF. G0A37039

The world's thinnest automatic-winding tourbillon owes its title to the house's 1270P caliber. Measuring a mere 10.4mm in thickness, this new interpretation of Piaget's aesthetic codes possesses an architecture as spectacular as the exposed movement that powers the watch's heartbeat. Through a laser- engraved sapphire crystal, the wearer may admire an offset white-gold micro-rotor lined with 107 brilliant cut diamonds (0.2 carat) as well as an imposing tourbillon cage, offering a veritable third dimension to a dial that features a delicate laser-guilloché craftsmanship and beautifully tapered hour markers. The case on this piece makes the model not only remarkable for watch afficianodos but for haute jewelry enthusiasts as well. The 18K white-gold case is set with over 1,000 diamonds weighing in at approximately 16.5 carats.

RICHARD MILLE

RM 021 TOURBILLON 'AERODYNE'

Continuing the expansion and application of truly unique materials to watchmaking, the RM 021 Tourbillon 'Aerodyne' is the first watch created with a composite baseplate utilizing a titanium exterior framework in combination with honeycombed orthorhombic titanium alumide and carbon nanofiber. This material was originally the subject of research by NASA for application as a core material of supersonic aircraft wings. This new tourbillion provides hours, minutes, indicators for power reserve (circa 70 hours), torque indicator and function indicator for winding, neutral, and hand-setting operations. The PVD-treated variable inertia balance wheel with overcoil spring and ceramic endstone for the tourbillon cage offers optimal chronometric results and excellent long-term wear capability.

RICHARD MILLE

RM 27-01 TOURBILLON RAFAEL NADAL

Inspired by the forces at play during high-level tennis matches, this new tourbillon in the collection represents an innovative development in the creation of tourbillon watches for the world of sports. The entire skeletonized movement of LITAL® alloy, weighing a mere 3.5 g in total, is attached to the monoblock case of carbon nanotube injected composite with 4 cables of just 0.35mm diameter, each with a uniquely designed attachment system that literally holds the movement suspended in the air. The case of the 50 pieces created for this limited edition also sports new visual cues in its tonneau case design: the front and back bezels extend to the surround areas of each case screw, giving an extreme sports look to the dynamic case lines whilst providing additional stiffness to the super-light case structure.

RICHARD MILLE

RM 036 TOURBILLON G SENSOR JEAN TODT

This new watch for 2013 unites the qualities of a tourbillon movement with a G-sensor indicator placed below 12 o'clock. The RM036 caliber, a tourbillon movement made entirely of grade-5 titanium and ARCAP, uses a carbon-nanofiber baseplate. This combination of materials provides the movement with optimal resistance and flatness, and draws a sublime contrast with the finishing that includes polishing, shot-blasting and satin-brushing. This new G-sensor system was developed and patented by APR&P. The G-sensor itself, composed of over 50 parts and measuring a mere 17mm, can withstand decelerations of several tens of G's. A percentage of the profits from the sale of this exceptional watch will be donated by Jean Todt to the ICM (Institut du Cerveau et de la Moelle Épinière – Brain and Spine Institute) and to the FIA Action For Road Safety, whose goal is to support the United Nations' ten-year action plan for road safety (2011-2020).

RICHARD MILLE

RM 56-01 TOURBILLON SAPPHIRE

The remarkable transparency of this tourbillon wristwatch starts with a technical feat that takes three months of work in its realization, with the machines that sculpt its exceptional contours working day and night, seven days a week. This new timepiece is a dais of brilliance, bringing light directly into the heart of the movement with a baseplate and bridges created from sapphire, one of the hardest materials known to man—only diamond is more resistant. The creation of such delicate movement parts in sapphire pushes the limits of horology as well as the engineering of materials to the outer limits of the possible. The result is unique in the world of timekeeping, creating a theater of scintillating transparency where prismatic light meets the mechanical mystery of the tourbillon.

RICHARD MILLE

RM 59-01 TOURBILLON YOHAN BLAKE

The 24-year-old Jamaican Yohan Blake, multiple medalist at the London Olympics in 2012, is a member of the very exclusive circle of the world's fastest sprinters and a partner of the brand. The RM 59-01 created for him has visually impressive, dynamically shaped bridges colored in green and yellow, machined from anticorodal aluminum Pb109, an alloy of aluminum, magnesium, silicon and lead. The skeletonized, manual-winding 21,600 vph tourbillon movement has a power reserve of approximately 50 hours. It offers an optimal performance/regularity/power reserve ratio, provided by the combination of grade-5 titanium with anticorodal aluminum Pb109, optimized gear trains and a variable-inertia balance. The asymmetrical case design is made from a translucent composite injected with millions of nanotubes in carbon, one of the strongest materials currently known to man.

RJ-ROMAIN JEROME

RED CHRONO TOURBILLON – REF. RJ.T.TO.CH.002.01

With its rugged good looks and adventurous spirit, the Chrono Tourbillon is the new flagship model of the Titanic-DNA collection by RJ-Romain Jerome. The secrets of its motorization are revealed through the openworked dial: each tooth and each intricate component of the chronograph and tourbillon are immediately visible. The bezel is made of oxidized and stabilized steel stemming from the extraordinary fusion of materials from the Titanic recovered from the depths of the Atlantic and steel from the Harland & Wolff shipyard in Belfast where the Titanic was built a century ago. It is also available in steel and black PVD-coated steel.

RJ-ROMAIN JEROME

STEAMPUNK BLACK TOURBILLON – REF. RJ.T.TO.SP.001.01

The Steampunk Black Tourbillon combines and galvanizes all the attributes of a devastating personality. Boasting polished steel paws, pistons and Roman numerals, this entirely dial-free timepiece features handmade rough finishing. It is housed within a Ø 50mm case that provides unobstructed views of the sturdy, rigorously constructed mechanism. Faithful to the Titanic-DNA collection, the bezel is in oxidized and stabilized steel, resulting from the fusion between the metal of the legendary wreck lying 3,840m below the sea, and that provided by the Harland & Wolff shipyard, birthplace of the Titanic.

ROGER DUBUIS

EXCALIBUR SKELETON DOUBLE FLYING TOURBILLON – REF. RDDBEX0281

As though the extraordinary creation of a double flying tourbillon with differential weren't enough, Roger Dubuis bares the thrill of this timepiece's structure with a skeleton design of the 319-component RD01SQ caliber. Carefully adjusted in six positions, this revealing masterpiece boasts two one-minute tourbillons, each with a frequency of 21,600 vph. Within the 45mm white-gold case and charcoal gray flange, the owner is treated to the intricacy of detail of the Roger Dubuis manufacture. This skeletal triumph of modern haute horology is secured to the wrist by a fully adjustable hand-stitched black alligator strap with white-gold folding buckle.

ROGER DUBUIS

LA MONEGASQUE FLYING TOURBILLON – REF. RDDBMG0010

Through the gold-plated aperture at 7:30, the golden tone of the gravity-defying flying tourbillon oscillates in powerful contrast to the assertive shimmer of the 42mm pink-gold case. With its off-centered positioning, the one-minute tourbillon shares its spotlight with the snailed silver power reserve at 4:00 and large date at 12:00 on the silver satin sunburst dial center. On the charcoal satin exterior, discreet Arabic numerals echo the gold color of this emblematic expression of haute horology. The RD540 that powers this timepiece grants it a power reserve of 60 hours. The Monegasque Flying Tourbillon is worn on a hand-stitched black alligator leather strap.

TAG HEUER

CARRERA MIKROTOURBILLONS – REF. CAR5A51.FC6323

The Carrera MikrotourbillonS is not only the world's fastest tourbillon: it is the first ever tourbillon on a 1/100th second chronograph that can be started and stopped. The MikrotourbillonS has two rotating tourbillon mechanisms visible on its dial face. The first beats at 4 Hz (28,800 vph) and controls the watch. The second, the world's fastest tourbillon, controls the 1/100th second chronograph and runs at 50 Hz, meaning it beats at 360,000 vph and rotates at a dizzying five seconds per revolution, or 12 times a minute. Another technical feat: the cage-free mechanism can be started and stopped thanks to its dual-chain architecture, which incorporates one kinematic chain for the watch and a second for the chronograph.

ZENITH

CAPTAIN TOURBILLON – REF. 18.2190.4041/01.C498

The tourbillon is a fascinating horological complication that finds its rightful place in this Captain model. A dial opening at 11:00 reveals the tourbillon rotating at a speed of one turn per minute. Equipped with El Primero Calibre 4041 S beating at the rhythm of 36,000 vph, this Captain Tourbillon owes its fascination both to the extraordinary expertise required to produce it and to its classic, timeless appearance that has become an enduring Zenith signature feature.

ZENITH

EL PRIMERO TOURBILLON – REF. 65.2050.4035/91.C714

Embodying an original association between a tourbillon and the legendary El Primero caliber, this timepiece comes in a 44mm case. It reveals an exceptional horological construction dedicated to serving a highly reliable movement, Calibre 4035 D, equipped with a tourbillon and a high-frequency chronograph. The tourbillon reveals its rotating carriage in the upper left-hand part of the dial, while incorporating a patented date display featuring a disc appearing around the periphery of the carriage.

Chronographs, Split-Seconds and Flyback Chronographs

TAG Heuer
Mikrogirder

Ricardo Guadalupe

CEO of Hublot

"Vertical integration and our manufacture are the watchwords at Hublot."

SINCE THE START OF 2012, Ricardo Guadalupe has been the new Hublot CEO. Having taken over from his friend Jean-Claude Biver, who is now Chairman of the Board of Directors, his objectives for 2013 include 15 new boutiques world-

On what note did 2012 end for Hublot?

Each of the past six years has been a record year for us, and 2012 was no exception. We are continuing to enjoy double-digit growth well above the average increase in Swiss exports as a whole and are currently achieving substantial turnover. This means it won't necessarily be possible to double this figure over the next three years. Hublot is now in a consolidation phase: we are really trying to establish the brand, notably through boutique openings, as well as through the growing development of our manufacturing facilities. Hublot is engaged in a process of vertical integration in terms of both movements and materials.

Speaking of integration, Hublot already has its own Unico chronograph movement. What proportion of in-house made movements can we expect to see in the future?

Hublot intends to become a must-have player on the Swiss watchmaking scene. Our ambition is to be quoted right up there among the reference brands such as Audemars Piguet, Jaeger-LeCoultre and Breguet. To achieve this goal, vertical integration and a full-fledged "manufacture" are the watchwords. In 2012, we produced 5,000 Unico movements, compared to around 35,000 watches produced in all. Within three to four years, the aim is to raise that figure to 20,000 Unico movements per year. That entails investments, notably in a new 6,000 square-meter manufacturing facility that will enable us to double our production area. This new building, which should be completed by the end of 2014, will be assigned to the Unico calibers as well as to materials such as ceramics, carbon fiber and Magic Gold (a scratchproof patented 18-karat gold).

UNICO MOVEMENT

Speaking of movements, does Hublot plan to develop an in-house automatic base movement?

We will continue with our current partnerships. We cooperate with Sellita for our base movement. Our working relations are fine and we plan to carry on this way. We also have a manufacture caliber, our ultra-thin hand-wound movement developed in partnership with La Joux-Perret. Hublot was involved right from the start in the development process. In future, we will thus pursue this strategy of a partnership with a manufacturer, but on a movement exclusive to Hublot.

What's the outlook for 2013? Will it be a year geared towards grand complications, chronographs, men's or women's models?

Our Masterpieces line is a creative think tank. For this collection we dream up developments with no predefined technical or aesthetic framework. In 2013, we are going to present a model stemming from a partnership with Ferrari, relating to the new Ferrari Enzo: the very peak of the range in the field of automobiles. The Ferrari and Hublot design teams cooperated in developing a product linked to this car. This Masterpiece Enzo model will be unveiled in two phases. First of all at the Geneva Car Show at the same time as the Ferrari, and then at the Hublot booth at BaselWorld. The product will of course be resolutely car-oriented, with a very long power reserve. We are indeed hoping to set a world record in this respect. It will be a very high-end watch with a price point of around 300,000 CHF. The movement, with its engine-inspired design, will be visible on the dial side. We want to succeed where others have failed in this partnership with Ferrari. As far as our Unico movement is concerned, our Big Bang model will be equipped with this caliber as of 2013. It was logical that we should decide to use our proprietary movement to power our icon watch. And of course, women will also be in the spotlight in 2013. They account for 30% of our sales and we regularly adapt existing models specifically for them.

Which are the growth markets? How many boutiques have you opened recently?

In 2012, we opened around 15 boutiques (58 Hublot boutiques at the end of 2012). And we'll be opening around 15 more in 2013, mostly in China, because that's where we need to progress. Greater China currently accounts for a little over 30% of watch exports, whereas Hublot only exports 6% of its production to this region. There are several reasons for that: we are a young brand that entered the Chinese market just three years ago. Our concept is very innovative and doesn't fit Chinese customers' standard expectations. Hublot wants to assert itself in China in accordance with its true nature—there is no question of giving the brand an artificial "Chinese touch" by offering specific models. The new generation of Chinese consumers want something different, a demand that Hublot is well equipped to meet. We are therefore going to tackle China by demonstrating, through a boutique network, that the brand can also establish itself there. Owning boutiques is fundamental in enabling end consumers to experience not just a product but also a whole world, a philosophy. That is extremely important for a brand such as ours.

In light of this determination to make a stronger impact on the Chinese market, the partnership with Ferrari appears to be even more meaningful…

Absolutely, because we need a stronger image. Chinese people like to buy a brand name. They therefore need to be reassured regarding the strength of the brand, and that is exactly what we are now working on. First of all by genuinely establishing the brand and then by showing who we are through our partnerships, notably the Ferrari partnership. Success in China will follow naturally.

RICARDO GUADALUPE (right) with
EDWIN FENECH (Ferrari Greater China President and CEO)

BIG BANG FERRARI

and dial, to a chronograph equipped with a TAG Heuer Calibre 1887 and a Cortech and Artecad case and dial—meaning entirely in-house made. In purchasing a Carrera watch, the consumer is acquiring a Manufacture-made watch at a price ranging from 3,800 to 5,800 CHF. Moreover, since our industrial processes are all established in Switzerland, it is a true Swiss Made model.

For Calibre 1887, we have also created a Carrera 50th anniversary Jack Heuer Edition, because we simply could not reserve the exceptional design of the Mikrogirder for a high-priced watch only. We thus decided to incorporate it within a 45mm diameter, instead of 48mm. In terms of the power reserve, this is a watch with greater stamina. The dial was designed by Jack Heuer and reveals his fascination for sports timers. It offers a nod to both the 19th century with the stopwatch function and to the 21st century with its extremely ergonomic pushers at 11 and 1 o'clock. Available at the price of 7,000 CHF, it is a non-numbered model bearing the production year.

Are the new Carrera models very different from the previous ones?

The Carrera is in fact evolving around Calibre 1887 in step with its steadily increasing production. We began two years ago with a 41mm Carrera in a production run of several thousand units. equipping our global bestseller with Calibre 1887 demonstrates our substantial in-house chronograph production capacity. The result is that, placed side to side, the two models look very similar, and yet everything has changed—including the wristband and the case, because when you change the movement, you must also change the case. We have also gone from an aluminum bezel to a ceramic one. So actually, while the new Carrera watches resemble the previous models, the two versions don't have a single screw in common!

CARRERA JACK HEUER EDITION

CARRERA CALIBRE 1887 CERAMIC BEZEL

Will the Carrera models be equipped in the future with in-house movements only?

Definitely not. We will always maintain certain Carrera range-entry models equipped with a Sellita SW500 or a Valjoux movement, even though most of the new developments are to be equipped with the 1887; with Calibre 1888 to be introduced this year; or else with Calibre 36 that already equips our Monaco and our Grand Carrera. Meanwhile, we also have the Mikro movements for the haute horology models. The Carrera collection will thus remain extremely broad, encompassing models from 3,800 to 200,000 CHF, but with movements that are exceptional within their respective price categories.

What kind of progress have you achieved in the domain of new materials?

As far as materials are concerned, we are presenting a carbon that is new in terms of the bonding agent and the processes involved. It is more suited to watchmaking because it is dermatologically compatible. The epoxy carbons used up until now have been called into question because of the risk of allergies and skin irritations. That is doubtless not a problem for a car, since the driver's skin rarely comes into contact with the chassis, but it is more so for a watch that by definition is in more or less constant contact with the skin. The second disadvantage of an epoxy carbon is its sensitivity to UV rays, which makes it unstable and thus liable to change color. For the carbon we are presenting on our new Carbon Matrix Composite Concept Chronograph, we have used a polyamide bonding agent better commercially known as nylon. Contrary to epoxy, it has no corrosive properties and is insensitive to UV rays, which makes it enduringly stable. The value of this carbon is that it is totally suited to industrial production, non-allergenic and stands the test of time.

CALIBRE 1887

It this a new material or a well-known material applied to watchmaking for the first time?

It's a first application of this material in the watchmaking field, and this naturally involved difficulties in adapting it to these micro-components; just as the belts of the V4 caliber came from the microsurgical field and required substantial developments to use them in watchmaking. In this case, we are talking about a process involving placing the carbon fibers in an epoxy mold, smoothing them in a vacuum and then firing them in a furnace. When they emerge from the mold, virtually no retouching is required and the texture of this material reflects a handcrafted rather than an industrial approach. Given that the fibers are never exactly identical, they lend a unique touch to each watch. A model equipped with Calibre 1887 weighs 72g, compared with 125g for a standard Carrera model. This Carbon Matrix Composite Concept Chronograph will be commercialized at the end of the year at a price of around 12,000 CHF.

What about Calibre 1888?

We launched the Calibre 1888 project at the start of 2012. We literally started with a blank page, even though this project naturally benefited from all the lessons learned in developing Calibre 1887 and the Mikro movements. Although the latter are not intended for large-scale production, they were 100% in-house made chronograph movements for which absolutely everything had to be built, developed and approved. The Mikro experience was thus extremely useful, in parallel with that which we gained from making Calibre 1887. All this means that within four years, we have in fact acquired a relatively co mprehensive range of experience in terms of calibers thanks to these two movement families. For the 1888, we wanted to develop the ideal caliber with a view to industrial mass production. The goal is to achieve relatively low costs—for both production and maintenance—that guarantee this caliber a certain degree of versatility. Another distinctive feature for Calibre 1888 is the position of the counters at 3, 6 and 9 o'clock. Chronograph structures are normally either 3-6-9 or 6-9-12.

CARRERA CARBON MATRIX COMPOSITE CONCEPT CHRONOGRAPH

We are spoiled for choice with the latter thanks to the Calibre 1887, Sellita and Valjoux movements, whereas Calibre 1888 will enable us to round out the range of options with a 3-6-9 dsign by adding a slender movement at a competitive price.

When will Calibre 1888 be introduced?

The idea is to bring it to market by the end of 2013. We will by then have taken 24 months to create it—from the blank page to when it rolls off the production line. This is an ambitious and very tight schedule, especially since we have tried to incorporate within this movement all the essential and ideal elements required for a chronograph. We have managed to achieve this only because we have gathered together all the necessary competencies at TAG Heuer and allocated them to this project.

Start, stop, reset…

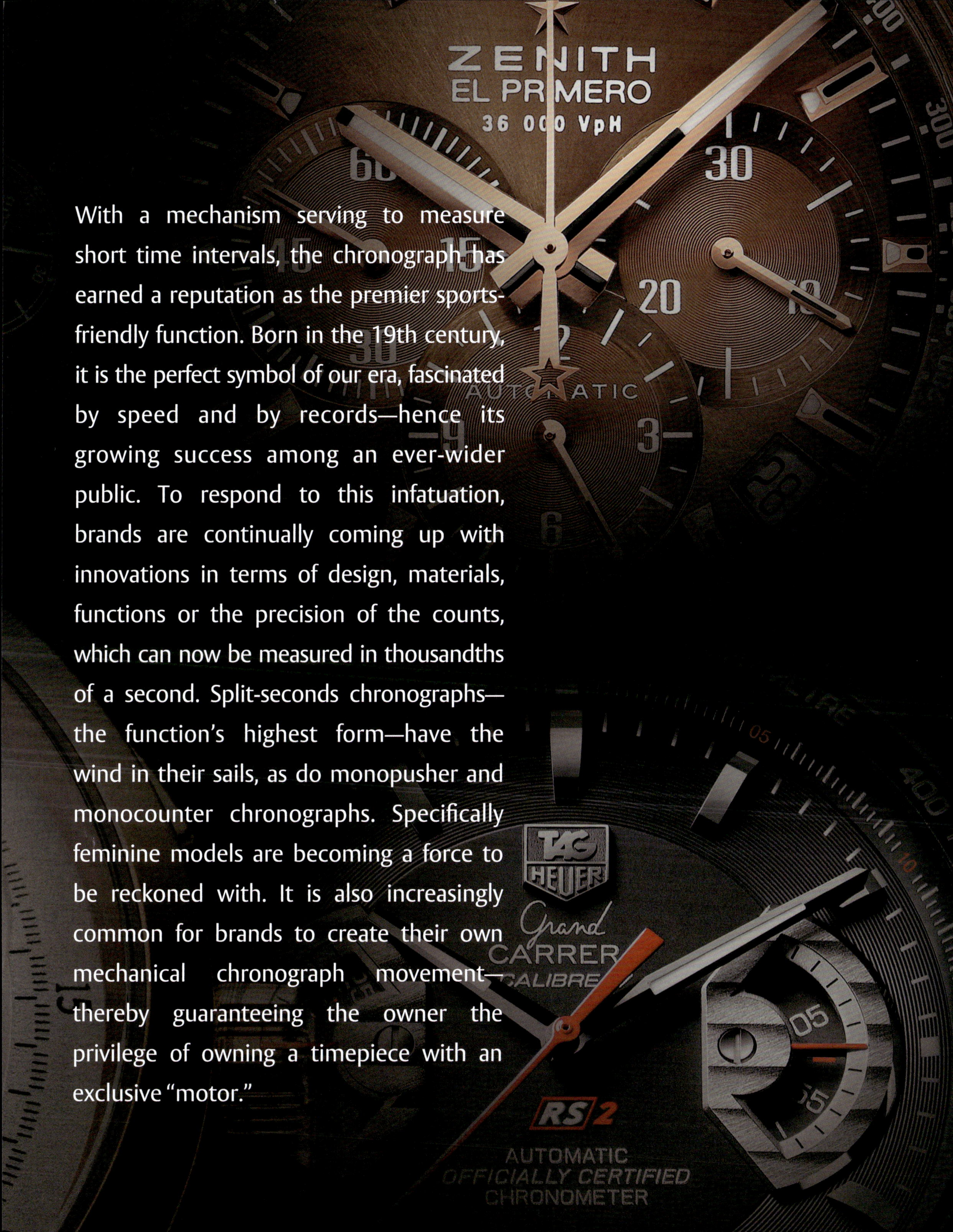

With a mechanism serving to measure short time intervals, the chronograph has earned a reputation as the premier sports-friendly function. Born in the 19th century, it is the perfect symbol of our era, fascinated by speed and by records—hence its growing success among an ever-wider public. To respond to this infatuation, brands are continually coming up with innovations in terms of design, materials, functions or the precision of the counts, which can now be measured in thousandths of a second. Split-seconds chronographs—the function's highest form—have the wind in their sails, as do monopusher and monocounter chronographs. Specifically feminine models are becoming a force to be reckoned with. It is also increasingly common for brands to create their own mechanical chronograph movement—thereby guaranteeing the owner the privilege of owning a timepiece with an exclusive "motor."

In 1821, **Nicolas Rieussec**—then "watchmaker to the King"—was tasked with timing a horse race on the Champs de Mars in Paris. He developed for the occasion a timepiece that he would patent the following year, under the name of "chronographe à seconde" (seconds chronograph). Accurate to one fifth of a second, this instrument was equipped with a tip that deposited a drop of ink on the dial at the beginning and at the end of each measurement. This invention would be perfected by Frederick Louis Fatton, who would transform it into a pocket chronograph with a fixed dial for Abraham-Louis Breguet. The year 1838 saw the invention of the split-seconds chronograph (see "How does it work?"), and the first system to reset the hands to zero came into being around mid-century. Now equipped with its three main functions (start, stop, reset), the chronograph would become very useful in several domains (science, technology, transportation, the military), and it is not by chance that its rise coincided with that of the industrial society and competitive athletics. During this period the chronograph would also gain several new scales for various measurements: tachometer (speed), telemeter (distance), pulsometer (pulse), etc.

The first wristwatch chronographs made their appearance in 1915. They were equipped, as were the pocket chronographs, with a single pushbutton for all three functions, usually housed in the crown. Breitling designed the now dominant chronograph configuration by adding the first independent pushbutton, and then in 1934, another one for resetting the chronograph. In 1937, the Dubois Dépraz manufacture presented a movement whose chronograph command operated using a cam system, less expensive and less difficult to produce than the traditional column wheel. The system would slowly be adopted by most brands.

In July of 1969, the Speedmaster, from Omega, would accompany American astronauts on their first visit to the moon. That year was also marked by the appearance of the first automatic chronograph movements. The movement developed by the Büren/Breitling/Heuer-Leonidas threesome in collaboration with Dubois-Dépraz was a cam movement featuring a modular construction, with an off-centered microrotor (oscillating weight); the one from Zenith, the famous El Primero, was an integrated movement with a column wheel in its central rotor. Thanks to its frequency of 36,000 vph, it was also the first chronograph movement capable of measuring times to one-tenth of a second.

The advent of quartz movements, which were much more precise, was nearly the deathblow for the mechanical chronograph, but the renaissance of mechanical horology in the late 1980s signaled its triumphant return. Let us remember that it is very rare for watch brands to design and produce their own chronograph movements. A large percentage of the chronographs on the market today house an ETA 7750 Valjoux movement, reworked to a greater or lesser degree and sold with differences in price that can be hard to justify... But the situation is beginning to change. The public, better informed on the subject than previously, is more than willing to pay a premium for a high-end chronograph, but on the condition that it is an exclusive movement—or at least somewhat original. The Swatch Group's announcement that it would stop selling ETA movements has also inspired competing brands to look for alternative solutions. For several years, we have been seeing an increase in the number of "in-house" chronographs featuring interesting innovations.

HOW DOES IT WORK?

The chronograph functions using a complex system of levers, springs and wheels. The chronograph mechanism is "grafted" onto the movement's base. When the "start" pushbutton is pressed, the chronograph wheel joins the wheel for the seconds, and the hand starts ticking. When the chronograph is stopped, the chronograph wheel separates from that of the seconds, and the chronograph hand stops again. The reset function is effectuated by a system of a hammer on a spring and a "heart."

SPLIT-SECONDS CHRONOGRAPHS

Split-seconds chronographs are equipped with an additional seconds hand, which can be stopped independently to measure a "split" time or a reference time. They are recognizable by the additional push-button they bear. Activating the pushbutton is all that is needed for the split-seconds hand to catch up with the main chronograph hand, so that the two may continue running together.

COLUMN-WHEEL AND CAM MOVEMENTS

There are two kinds of chronographs with respect to the "control group" governing the start, stop and reset functions. The chronographs considered to be the most prestigious are equipped with a column-wheel, a complex organ that uses a system of slots and tabs to govern the movements of the levers. Most current mechanical chronographs are equipped with a cam system of simpler construction.

MODULAR AND INTEGRATED CONSTRUCTIONS

In modular constructions, the chronograph mechanism is totally independent of the base movement. It is mounted on an additional plaque affixed to the dial side of the movement. This solution allows the watchmaker to link a chronograph mechanism to an existing movement, without having to develop a new caliber. In integrated constructions, the chronograph components are placed directly on the movement, on the bridge side. Column-wheel chronograph movements are the result of integrated constructions.

FREQUENCY AND PRECISION

The precision of time measurement and display of a mechanical chronograph depend on the frequency of its regulating organ, i.e. the number of vibrations per hour (one oscillation equals two vibrations) of the balance-spring. The higher the frequency, the smaller units of time that can be measured.

Vibrations per hour	Frequency (in Hz)	Smallest measurable unit of time
18,000	2.5 Hz	1/5 of a second
21,600	3 Hz	1/6 of a second
28,800	4 Hz	1/8 of a second
36,000	5 Hz	1/10 of a second
360,000	50 Hz	1/100 of a second

FLYBACK FUNCTION

The flyback function allows the wearer to instantly start a new round of timing while the chronograph is already operating. For this, one need only activate the reset pushbutton, without stopping the hands first; all three chronograph hands will immediately return to zero to start another measurement.

JUMPING SECONDS

The flyback function allows the wearer to instantly start a new round of timing while the chronograph is already operating. For this, one need only activate the reset pushbutton, without stopping the hands first; all three chronograph hands will immediately return to zero to start another measurement.

CHRONOGRAPH AND CHRONOMETER

The word "chronometer" is often used incorrectly to designate a chronograph, but they are two very different things. A chronometer is a watch whose movement has received an official certificate attesting to its extreme precision. This certificate is awarded by the COSC (Swiss Official Chronometer Testing Institute) after two weeks of very rigorous testing. A chronograph may be a certified chronometer, but it is by no means certain that any given chronometer will also be a chronograph.

A Reinvented Classicism

Chronographs have long been distinguished by their rather classic aesthetic, with three small symmetrically placed subdials. This traditional look has now been swept aside by many models that reinvent the ways in which we measure time. The chronograph universe has also been marked by the appearance of new materials that allow the watches to become lighter, more robust or more technically sophisticated, as well as shaking up aesthetic codes.

Omega, *Speedmaster Professional*

The only watch to have been worn on the moon—hence its Moonwatch nickname—the famous Omega Speedmaster is presented in a version that stays true to the original, recognizable for its ultra-readable black dial and its tachometer.

Zenith, *El Primero Chronomaster 1969*

Equipped with the famous El Primero caliber, Zenith's El Primero Chronomaster 1969 takes on the historic color scheme of the first automatic-winding chronograph with integrated movement (1969). An aperture on the dial reveals the regulating organ, oscillating at 36,000 vph.

Rolex, *Cosmograph Daytona*

Created in 1963 for racecar drivers, Rolex's Cosmograph Daytona has become one of the most legendary chronograph models. It is equipped with a completely Rolex-made movement with an ultra-shock-resistant Parachron balance spring.

Piaget, *Gouverneur Chronograph*

Piaget has equipped its Gouverneur Chronograph with a new, house-made, ultra-thin, automatic-winding movement with flyback function. This two-counter chronograph (seconds/minutes) also displays a second time zone on a 24-hour scale at 9:00.

Vacheron Constantin
Overseas Chronograph

The Overseas Chronograph's bezel and bracelet design recalls the famous Maltese cross, Vacheron Constantin's emblem. The movement is girded by a metal ring that protects it entirely from magnetic fields.

Chopard, *Superfast Chrono*

On its Superfast Chrono, Chopard joins an ultra-sporty, automotive racing-inspired look to an automatic chronograph manufacture caliber, equipped with a flyback function.

Breitling, *Chronomat 44*

Breitling's flagship model, the Chronomat received a new movement in 2009—an automatic chronograph movement entirely developed and produced in-house. Like all the brand's movements, it is a certified chronometer.

Quinting, *Mysterious Chrono*

To create the Mysterious Chrono, which offers a magnificent take on transparency, Quinting developed a 237-component mechanical movement powered by four motors hidden within the circumference of the watch.

Eberhard & Co., *Chrono 4 Géant*

Equipped with four horizontally aligned counters, the Chrono 4 Géant from Eberhard & Co. is available in a titanium limited edition of just 1,887 pieces: an homage to the year in which the house was founded.

Jaeger-LeCoultre
AMVOX2 Chronograph Concept

On its AMVOX2 Chronograph Concept watch, Jaeger-LeCoultre introduced a small revolution in terms of controls. The traditional pushbuttons were replaced by a system in which the case middle pivots around a horizontal axis from 9 to 3 o'clock. To start or stop the chronograph, one need only press gently on the upper half of the crystal; resetting is done by a push on the lower half.

Hanhart
Pioneer Stealth 1882 Limited Edition

On the limited edition Pioneer Stealth 1882, created for the brand's 130th anniversary, Hanhart picks up several details that distinguished its World War II aviator chronographs—such as the counters placed at the exterior of the dial, an hour totalizer integrated with the small seconds display and a red reset pushbutton.

Dubey & Schaldenbrand
Grand Dôme DT

On the tonneau-shaped Grand Dôme DT from Dubey & Schaldenbrand, the chronograph displays are arranged along a vertical axis, with minutes at 12 o'clock and hours at 6 o'clock.

Audemars Piguet, *Millenary MC12 Tourbillon Chronograph*

The Millenary MC12 Tourbillon Chronograph from Audemars Piguet combines high-tech materials (carbon mainplate, anodized aluminum bridges) with a daring automotive-inspired design.

Montblanc, *Metamorphosis*

On the spectacular Metamorphosis from Montblanc, a bolt at 9 o'clock causes the hours/minutes/seconds/date dial to disappear as if by magic, making way for the chronograph's minute counter, as the central seconds hand becomes that of the chronograph function.

Disc Displays

Varying the looks and methods of reading the time, watchmakers sometimes have a little fun by replacing the chronograph minutes hand (or even the hour or seconds counter hand) with an aperture and disc display—an excellent way to combine originality and readability.

Cartier

Rotonde Central Chronograph

On the Rotonde Central Chronograph from Cartier, the chronograph indications have the place of honor at the center of the dial, with a seconds hand and a counter that sweeps across an arc as a 30-minute totalizer.

Jaeger-LeCoultre
Master Compressor Extreme LAB 2

Designed to withstand even the most challenging conditions, the Jaeger-LeCoultre Master Compressor Extreme LAB 2 is distinguished by its patented digital counter and its very readable way of indicating the chronograph's jumping minutes.

Pierre DeRoche
GrandCliff TNT Penta

Equipped with an exclusive Dubois Dépraz movement, the GrandCliff TNT Penta uses five discs (hence its name) to display the large date and the small seconds as well as the 60-minute and 12-hour counters of the flyback chronograph.

TAG Heuer
Grand Carrera Calibre 17 RS2

The Grand Carrera Calibre 17 RS2 from TAG Heuer utilizes its Rotating System to present two disc-type indications: small seconds at 3 o'clock and chronograph minute counter at 9 o'clock.

Porsche Design, *P'6910 Indicator*

On the P'6910 Indicator chronograph from Porsche Design—with a resolutely automotive look—the hours and minutes of the chronograph function are displayed "digitally" with discs rotating past a double aperture at 3 o'clock.

Montblanc
Nicolas Rieussec Open Date Silicium Escapement

A tribute to the inventor of the first chronograph, the Nicolas Rieussec Open Date Silicium Escapement from Montblanc displays the chronograph minutes and seconds using two rotating discs equipped with fixed pointers.

Time for Women

A sign of the times, and a notable trend at the beginning of the third millennium, a growing number of chronographs are designed exclusively for feminine wrists. These sporty, elegant watches are often adorned in soft colors and graced with finely finished dials featuring materials such as mother-of-pearl and diamonds, following the lead of more formal watches. These beauties meld technical excellence and aesthetic refinement.

Zenith
El Primero 36,000 vph 38mm Lady
In its elegant, rose-gold, diamond-set case, paired with a chocolate-brown dial, Zenith's El Primero 36,000 vph 38mm Lady houses the famous automatic-winding chronograph caliber, precise to one tenth of a second.

Girard-Perregaux
Small Column-Wheel Chronograph
Girard-Perregaux offers a beautiful combination of technique and aesthetics on its automatic-winding Small Column-Wheel Chronograph for women, whose openworked dial features four large Arabic numerals of highly original design.

Patek Philippe
Ladies First Chronograph
Ref. 7071
Launched in 2009, the Ladies First Chronograph Ref. 7071 from Patek Philippe was the first watch to house the brand-new in-house chronograph caliber CH 29-535 PS, which is classically constructed; it has since appeared in a masculine model.

Harry Winston
Premier Large Chronograph
Equipped with a sunray dial and a diamond setting on the bezel and lugs, the Premier Large Chronograph from Harry Winston unites a sophisticated technique with the characteristic refinement of Harry Winston jewelry.

The Fashion for Single Counters

Eschewing the system of two small separate counters for the minutes and the hours, a growing number of watchmakers are opting for a single counter that displays both indications. The goal is to enable simpler, quicker reading of longer time measurements, just as one might read the time on the central dial.

Pierre DeRoche, *SplitRock*

On its SplitRock, Pierre DeRoche emphasizes simplicity and ease of reading by proposing a chronograph with three concentric hour/minute/second hands, equipped with an exclusive Dubois Dépraz caliber.

IWC
Portuguese Watch Club Chronograph

A new interpretation of an iconic IWC model, the Portuguese Watch Club Chronograph displays the chronographs hours and minutes in a subdial at 12 o'clock. The chronograph also boasts a flyback function.

Jaeger-LeCoultre
Duomètre à chronographe

On the Duomètre à chronographe by Jaeger-LeCoultre, equipped with two independent mechanisms synchronized by the same regulating organ, the hour and minute counters are grouped together at 2:30, while a flying seconds hand at 6 o'clock allows measurements of one sixth of a second.

Omega
De Ville Co-Axial Chronograph

Offered in steel or red gold, Omega's De Ville Co-Axial Chronograph is distinguished by its single counter for hours and minutes at 3 o'clock, as well as by its movement with exclusive co-axial escapement.

Patek Philippe
Annual Calendar Chronograph Ref. 5960

The Annual Calendar Chronograph Ref. 5960 from Patek Philippe utilizes a vertical disc clutch system, which allows the watch to use the central hand for chronograph seconds and a single subdial for the minute and hour counters.

Single-Pusher Chronographs

In terms of specialty constructions, we are seeing an increase in single-pusher, or monopusher, chronographs. These pieces feature a sophisticated arrangement: the sole pushbutton handles the start, stop and reset functions, in succession.

Jaquet Droz
Complication La Chaux-de-Fonds Chrono Monopusher
The Complication La Chaux-de-Fonds Chrono Monopusher from Jaquet Droz is notable for being the first timepiece of its kind equipped with an hour counter as well as an off-centered hour and minute indication.

Blancpain, *Villeret Single-Pusher Chronograph*
Available in steel or red gold, the Villeret Single-Pusher Chronograph from Blancpain also displays a complete calendar—date indicated by a pointer, day, month and moonphase—upon its refined dial.

Christophe Claret
DualTow NightEagle
The DualTow model from Christophe Claret, shown here in its NightEagle version, is especially notable for its planetary-gear system, which drives the hours, minutes and seconds, and limits the variations in amplitude of the balance spring, providing a much greater precision.

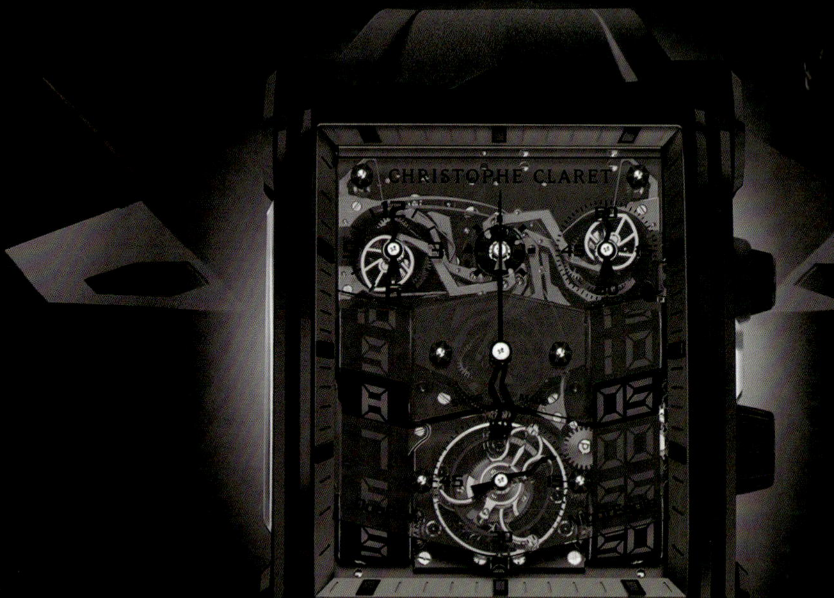

Montblanc, *Grand Chronographe Régulateur*

Montblanc confirms its rise through the ranks by unveiling the Grand Chronographe Régulateur, a single-pusher chronograph with an hour display at 12 o'clock and a large central minutes hand.

Corum, *Admiral's Cup Chronograph 44 Centro Mono-pusher*

The Admiral's Cup Chronograph 44 Centro Mono-pusher from Corum is distinguished by its highly readable system of a red-tipped central hand that acts as a minute counter for the chronograph.

Longines, *Column-Wheel Single Push-Piece Chronograph 180th Anniversary Limited Edition*

To celebrate its 180th anniversary, Longines launched a limited edition, monopusher, column-wheel chronograph. The design of this rose-gold piece pays tribute to the first chronograph wristwatch released by Longines in 1913.

Hundredth-of-a-second—or thousandth-of-a-second—timing

For many years, the display of time to one tenth of a second reigned as the supreme accomplishment for a mechanical chronograph. A few watchmakers have taken up the challenge, daring themselves to break the sound barrier and present chronographs able to measure and display one one-hundredth and even one one-thousandth, of a second—notably using bi-frequency systems, with a regulating organ for the watch part and a separate regulating organ for the chronograph part. It is an admirable technical advance, even though the inevitable reaction time and delay in activating the pushbutton limits the usefulness of such a function in the field.

Montblanc, *TimeWriter II Chronograph Bi-Fréquence 1000*

On its TimeWriter II Chronograph Bi-Fréquence 1000, Montblanc measures time to 1/1000 of a second with a second balance that "only" beats at 360,000 vph (50Hz). The "sweep seconds-hand" makes a complete revolution around the dial in one second, and the thousandths of a second are displayed in an aperture at 12 o'clock.

F.P. Journe, *Centigraphe Sport*

On the Centigraphe Sport from F.P. Journe, equipped with a single balance wheel beating at 21,600 vph, hundredths of a second are read off using a "flying" hand that makes a full rotation in one second. The tachometric scale can measure speeds of up to 360,000 km/h! Entirely crafted in aluminum alloy, this chronograph weighs just 55g.

TAG Heuer,
Carrera Mikrograph 1/100th Second Chronograph

To display the hundredth of a second using a central hand, the Carrera Mikrograph 1/100th Second Chronograph from TAG Heuer is equipped with two escapement systems in the same movement, of which one balance-wheel (powering the chronograph) oscillates at a frequency of 360,000 vph.

TAG Heuer, *Mikrogirder*

Equipped with an unusual regulating organ oscillating at 1,000Hz (7,200,000 vph), TAG Heuer's concept watch Mikrogirder measures time to 5/10,000 or 1/2,000 of a second. Its bi-frequency system uses two independent "chains," one for the watch and one for the chronograph.

Split-Seconds Chronographs

Split-seconds chronographs possess an additional seconds hand that can be stopped to measure an intermediate or split time. This ultra-sophisticated clutch system is widely recognized as one of the most difficult complications to master—the equal of a tourbillon or a minute repeater—and the most prestigious brands are eager to offer it among their collections.

A. Lange & Söhne, *Lange Double Split*

The German brand A. Lange & Söhne presents a new innovation with its Lange Double Split, the first wristwatch chronograph with both a split-seconds mechanism and a split-minutes mechanism.

Patek Philippe

Ladies First Split Second Chronograph Ref. 7059

Patek Philippe paid tribute to women with the launch of its Ladies First Split Second Chronograph Ref. 7059, a single-pusher split-seconds chronograph powered by an ultra-thin movement that also drives several men's models.

Blancpain

L-evolution Chronographe Flyback à Rattrapante

Joining carbon fibers and cutting-edge technology, the L-evolution Chronographe Flyback à Rattrapante reflects Blancpain's engagement in the world of automotive competition. It is also equipped with a two-aperture large date display.

Richard Mille, *RM 056 Tourbillon Split Seconds Competition Chronograph Felipe Massa Sapphire*

Dedicated to the Brazilian Formula 1 driver of the same name, Richard Mille's RM 056 Tourbillon Split Seconds Competition Chronograph Felipe Massa Sapphire is notably distinguished by its completely transparent case, carved from a solid block of sapphire.

Panerai

Radiomir Regatta 1/8th Second

The split-seconds chronograph Radiomir Regatta 1/8th Second possesses two counters: one with a flying second that makes a complete one-second rotation in increments of one eighth of a second, the other with a scale graduated in nautical knots that calculates the speed of a boat.

Regatta Chronographs

Models described as "regatta chronographs" possess a notable countdown system that marks off the last five or ten minutes before the race begins. With the growing popularity of sailing competitions such as the America's Cup, these models are getting new wind in their sails.

Glashütte Original, *PanoRetroGraph*

Though it is not a true regatta watch, the PanoRetroGraph from the German brand Glashütte Original stands out for its mechanical chronograph system that allows the wearer to count "up" as well as "down."

Louis Vuitton , *Tambour Régate America's Cup*

In Louis Vuitton's limited edition Tambour Régate America's Cup, launched for the 34th America's Cup in 2013, the usual round countdown apertures have been replaced by five trapezoidal cutouts that form the silhouette of a spinnaker.

Rolex, *Oyster Perpetual Yacht-Master II*

Rolex has reaffirmed its place in the sailing world by launching the Oyster Perpetual Yacht-Master II, which boasts an ingenious bezel/crown/pushbutton system that follows the countdown using a large triangular hand.

Alpina, *Extreme Sailing*

To mark its entry into the world of nautical sponsorship, Alpina created the Extreme Sailing regatta chronograph. The dial uses a traditional countdown display method, with the help of five circles that in turn change from blue to red.

A. LANGE & SÖHNE

DOUBLE SPLIT – REF. 404.032

This model is powered by the manual-winding Lange L001.1 movement. The Double Split is a flyback chronograph with double rattrapante, controlled by classic column wheels. Included is a jumping chrono minute counter and rattrapante minute counter, flyback function with indication for hours, minutes, small seconds with stop seconds and power reserve indicator. The movement is encased in a 43mm rose-gold case with sapphire crystal on the front and caseback. The dial is two-tiered solid silver. The Double Split is also available in platinum.

AUDEMARS PIGUET

ROYAL OAK OFFSHORE CHRONOGRAPH – REF. 26401PO.OO.A018CR.01

The Royal Oak Offshore 44mm is powered by Audemars Piguet's in-house, automatic-winding Manufacture caliber 3120, which is comprised of 280 parts and beats at 21,600 vph. The movement is enclosed in a 950 platinum case and includes a black ceramic bezel and crown, as well as pushbuttons and pushbutton guards. The chronograph features a date with the small seconds at 12:00. The pin buckle is also made of 950 platinum and the sapphire crystal caseback reveals the subtlety of its mechanism. This piece has anti-magnetic protection and is water resistant to 100m.

BELL & ROSS

WW1 CHRONOGRAPHE MONOPOUSSOIR IVORY DIAL

Paying tribute to timepieces donned by aviators during World War 1, the WW1 Chronographe Monopoussoir bears precise similarities to chronographs that were popular amongst pilots in the 1920s. With only one button to press and a sharp, decipherable dial to read, this chronograph assists with measuring short periods of time. Integrated into the polished-steel crown, the pushbutton controls the chronograph's functions: start, stop and reset. Bell & Ross, a master of military timekeeping, emphasizes the importance of legibility by updating the mechanical design with an increased distance between the 60-second and 30-minute counters.

BREGUET

CHRONOGRAPH TYPE XXII – REF. 3880ST.H2.3XV

A technical exploit centered on the expert use of silicon in the movement, this self-winding chronograph raises the bar for high-performance haute horology. Performing a full dial revolution in 30 seconds, the central chronograph hand offers twice the precision and accuracy of readability, all made possible by the 589F caliber's exceptional rate of oscillation of 72,000 vph. The first series-made chronograph with a 10Hz frequency, the 44mm steel timepiece is additionally endowed with a flyback function, date, secondary time zone and complementary central hand to clarify the chronograph's status within the first or second half of the minute.

CHRISTOPHE CLARET

DUALTOW – REF. MTR.CC20A.064

The DualTow, created to celebrate the 20th Anniversary of the Claret Manufacture, is a single-pusher planetary-gear chronograph with striking mechanism and tourbillon. The major innovation of this timepiece, issued in a limited edition of 68 unique and customizable watches, lies first and foremost in its patented, world premiere chronograph function. This remarkable chronograph operates 3 differential systems for the seconds, minutes and hours by means of 9 planetary gears all through a single pusher, which simultaneously activates a gong for each change of function (start, stop and reset)! Its exceptional belt-type display indicating the hours and minutes, bestows high-tech sophistication.

CHRISTOPHE CLARET

DUALTOW NIGHTEAGLE – REF. MTR.CC20A.053

The latest version of the DualTow model, named the DualTow NightEagle, features a variety of inspirations including the ultra secret world of stealth aircraft such as the American F-117 Night Hawk. It reflects these taut lines, sharp angles, tone-on-tone colors and transparency effects. The Grade-5 black PVD titanium hour and minute display on black Lipari belts features a dial with blue numerals. The upper part of the movement features three tinted sapphire bridges, made in house, which form a stealth looking dial. The NightEagle unique version is quite special as its total production is just 68 unique and customizable models.

CHRISTOPHE CLARET

KANTHAROS – REF. MTR.MBA13.905

A 2013 novelty, this automatic-winding model is powered by a MBA13 caliber with a power reserve of 40 hours. This new watch is comprised of 558 parts and beats at 21,600 vph. Complete with a monopusher chronograph as well as Christophe Claret's signature patented cathedral gong chime, the Kantharos exhibits fine watchmaking qualities throughout. The mechanical chime is activated each time the function is changed (start, stop, reset). On the dial at 6:00 the constant force mechanism is visible as well as underneath the sapphire bridge. It is available in several materials ranging from white gold to pink gold as well as titanium.

GLASHÜTTE ORIGINAL

PANOGRAPH – REF. 61-03-25-15-05

The PanoGraph uses a 40mm case and expansive dial to represent the Pano line's characteristic asymmetric visuals in a contemporary manner. The design arranges the interlocking hour/minute and seconds dials to the left, while Glashütte Original's panorama date display occupies its usual position at 4:00. Two chronograph pushbuttons at 4:00 and 2:00 are labeled START-STOP and FLY-BACK, respectively. Presented in a red-gold case, the PanoGraph visuals represent the epitome of simplicity and aesthetic harmony, based on the golden ratio. At the heart of this mechanical wonder is the Calibre 61-03, designed and hand-crafted in the Glashütte Original manufacture. The manually wound column-wheel chronograph is outfitted with a classic switching wheel that operates the various chronograph functions, including start, stop, reset and a satisfying flyback mechanism.

GUY ELLIA

JUMBO CHRONO

The Jumbo Chrono, the first round men's watch created by Guy Ellia, is based on an interesting technological approach demonstrating that it is possible to combine aesthetic balance with masculine detail and a fine mechanical movement. The Jumbo Chrono boasts an exceptional 50mm size that is rarely seen in the world of luxury watches. The Jumbo Chrono's rhodium-plated rotor is unique and the impressive discovery dial reflects a strong graphic design: sweep-seconds chronograph in the center, hour and minute at 12:00, date at 2:00, 30-minute counter at 3:00, seconds at 6:00, and 12-hour counter at 9:00. This model is shown in black gold, but is also available in white or pink gold.

HARRY WINSTON

HARRY WINSTON OCEAN DIVER – REF. 410/MCA44RZC.KR

A timepiece notable not only for its technical mastery but also for its stylish personality, the Harry Winston Ocean Diver is the ultimate watch for the individual who is willing to take risks in the water or on dry land. The unidirectional rotating bezel can be used for timing decompression stops, with highly legible and luminous hands and hour markers. This model is water resistant to 200m, an international requirement for diving watches. In addition to being a diver's watch, this piece is also a chronograph with a 30-minute counter at 9:00 and a 12-hour counter at 6:00. The dial presents practical information to its wearer within the framework of an ornamental and three-dimensional sculpture. This model's red accents and impeccable design make for a unique diving chronograph.

HARRY WINSTON

OCEAN SPORT™ CHRONOGRAPH – REF. 411/MCA44ZC.K

Self-winding with a skeletonized chronograph module, the Ocean Sport™ Chronograph is as sophisticated as it is functional. With a power reserve of 42 hours, the black ruthenium movement is with a satin-brushed finish. The dark, openworked three-dimensional dial of this sporty timepiece serves as the ultra-masculine background for its chronograph subdials. The 12-hour counter at 6:00 and 30-minute counter at 9:00 are accompanied by a date aperture at 12:00. The case, impeccably crafted in Harry Winston's ingenious and ultra-durable Zalium™, has a unidirectional rotating bezel with a satin finish and DLC-coated indexes.

HARRY WINSTON

OCEAN SPORT LADIES CHRONOGRAPH – REF. 411/LCA38ZC.AD/D02

Harry Winston's Ocean Sport Ladies Chronograph is the perfect watch for the sophisticated woman. This piece showcases the house's knack for haute horology as well as haute jewelry, as it is a diamond-set, technical masterpiece. The rhodium-colored spiral-shaped engraved dial has a sunray satin-brushed central finish. The off-centered silver-colored ring is carefully crafted with Côtes de Genève finishing and set with one brilliant-cut diamond in its center. The hours and minutes subdials, at 6:00 and 9:00 respectively, are snailed with rhodium-colored finishes. Blue hour markers are set with 15 brilliant-cut diamonds while the Arabic numeral at 12:00, hands and shuriken shine from SuperLumiNova coating. This beautiful face is framed by a Ø 38mm Zalium™ case and polished and notched engraved bezel that is carefully set with 9 brilliant-cut diamonds. The watch rests atop a white rubber strap that completes the strong and feminine look of the piece.

HUBLOT

BIG BANG BOA BANG – REF. 341.PX.7818.PR.1978

Bringing a luxurious taste of the jungle to the wrist, the Big Bang Boa Bang adorns its decisive reptilian theme with a vibrant collection of precious stones. The 41mm 18K red-gold case sets the stage for 48 baguette-cut light green tsavorites, tourmalines and light green sapphires. On the dial, the theme of the watch is confirmed with a green snakeskin print, invigorated by 8 diamonds acting as indexes. Worn on a green python snakeskin strap on black rubber, the Big Bang Boa Bang is powered by the self-winding HUB4300 chronograph movement and features the date via a trapezoidal aperture at 4:30.

HUBLOT

BIG BANG FERRARI – REF. 401.MX.0123.GR

This 45.5mm exhibition of horological innovation derives its power from the in-house HUB1241 Unico movement, a self-winding chronograph caliber with flyback capabilities and column wheel. The result of extensive and dedicated research and development, the case is made of Magic Gold, an extraordinary Hublot scratch-resistant 18K gold alloy. The Big Bang Ferrari is imbued with racing theme and pedigree, exhibited beyond the movement by the wheel-rim-shaped oscillating weight, engraved Ferrari logo on one of the two race-inspired micro-blasted black PVD titanium pushers, and the iconic Ferrari prancing horse, rhodium plated and visible at 9:00.

HUBLOT

KING POWER F1 AUSTIN – REF. 703.NQ.8512.HR.FTX12

Celebrating Formula 1's awaited return to the United States, in Austin, Texas, Hublot further declares itself a speed-driven manufacture with an innovative chronograph loaded with high-tech construction. Upon the 48mm satin-finished titanium case, a racing-inspired "brake-disc" style titanium and carbon fiber bezel is anchored by 6 H-shaped titanium screws. The decisive cockpit premise of the dial boasts a gray F1 logo at 12:00 and is brought to life by two sharp satin-finished hands enhanced with black SuperLumiNova. Limited to 250 individually numbered pieces and powered by the automatic HUB4100 chronograph movement, the King Power F1 Austin makes abundantly clear the house's stance on high-speed racing's place in America.

HUBLOT

KING POWER UNICO KING GOLD – REF. 701.OQ.0180.RX

Symbolic of Hublot's progress toward verticalization, this timepiece is proudly driven by the manufacture's first movement produced entirely in-house. The Unico movement, a 330-component column-wheel flyback chronograph with silicon escapement, oscillates at a frequency of 28,800 vph and boasts a calendar display at 4:30. With its case, caseback, and clasp in 18K King Gold, as well as hands and appliqués gold-plated with white luminescence, the King Power Unico King Gold undoubtedly earns its name. Adding a distinctive touch of powerful aesthetic, the bezel is constructed of carbon fiber and black rubber.

HUBLOT

KING POWER USAIN BOLT – REF. 703.CI.1129.NR.USB12

Encapsulating Hublot's thorough connection to high-speed racing, the King Power Usain Bolt puts the manufacture on the wrist of the world's fastest man. The automatic-winding chronograph, measuring 48mm in diameter, wittily injects the character of its muse in the very fabric of its construction. The micro-blasted black ceramic timepiece is worn on a gold-colored strap made of the exact same synthetic leather as the legendary sprinter's shoes. In addition, the silhouette of Bolt's iconic pose is instantly recognizable in anthracite gray within the small seconds counter at 9:00. Finally, a touch of green in the 12-hour chronograph subdial tips Hublot's cap to Usain Bolt's beloved home of Jamaica.

IWC

SPITFIRE CHRONOGRAPH – REF. IW387802

The Spitfire Chronograph timepieces from IWC are notable for their traditional charm and complex mechanical development. These classic timepieces have been reworked with a slate-gray dial; the light that falls onto the sun-patterned finish alludes to circular motion on the smoky dark surface. On this particular model, the sleek combination of the special dial, perfectly matched stainless steel case and chocolate brown alligator leather strap showcase a classic complication: the flyback chronograph. The minutes subdial at 12:00, central seconds hand and small seconds subdial at 6:00 allow the wearer to measure time to the most precise interval. The Spitfire chronograph is also available with an equally luxurious 18K red-gold case.

IWC

PILOT'S WATCH CHRONOGRAPH TOP GUN MIRAMAR – REF. IW388002

Powered by the self-winding 89365 caliber, which comprises 35 jewels and provides 68 hours of power reserve, this watch is the quintessential military timepiece. The olive green hue of the strap, metallic undertones on the dial and black ceramic case all pay homage to the army. The unique dial's chapter ring and inner hour circle alludes to deck watches from the 1930s and 1940s. The Pilot's Watch Chronograph Top Gun is also available in black (Ref. IW388002) and both models are complete with a flyback chronograph function as well as protection against magnetic fields.

IWC

PILOT'S WATCH CHRONOGRAPH – REF. IW377701

The ultimate timepiece for aviators, the Pilot's Watch Chronograph is not only precise, but also extremely functional. The stainless steel case and instrumental matte black dial showcase a triple date display window at 3:00 along with a triangle at 12:00 and bright white luminescent hands that optimize legibility for pilots. This piece also contains a chronograph with a 12-hour counter at 6:00, a 30-minute counter at 12:00 and split-seconds central hands. The chronograph and the small seconds subdial at 6:00 provide promise precision for the wearer. This model houses an automatic-winding 79320 caliber with a power reserve of 44 hours, and beats at 28,800 vph.

LONGINES

CONQUEST CLASSIC – REF. L2.786.8.76.3

Conquest Classic joins the brand's collections of classical beauty that have helped to make Longines's reputation and success throughout the world. With a diameter of 41mm, this chronograph in rose gold, fitted with a black alligator strap, houses the exclusive mechanical column-wheel caliber L688. Its silvered dial is set with one Arabic numeral and 11 applied indexes with SuperLumiNova. It displays the hours and the minutes, as well as small seconds at 9:00, a 30-minute counter at 3:00, a 12-hour counter at 6:00 and the date at 4:30. Just like the entire Conquest Classic collection, this timepiece is water resistant to 5 bar and its screw-down caseback is fitted with sapphire crystal.

PANERAI

LUMINOR 1950 RATTRAPANTE 8 DAYS TITANIO – PANERAI TRANSAT CLASSIQUE 2012 REF. PAM00427

Celebrating Panerai's vintage-yacht heritage and tradition, this brushed-titanium timepiece, with rattrapante hand for the simultaneous measure of a complementary event to the chronograph function, is adorned with a caseback engraving of the Panerai Transat Classique's logo. Honoring a spectacular and distinguished race, the 47mm timepiece's eight-day power reserve can be read on a linear gauge at 6:00. Powered by a cutting-edge movement with vertical clutch and twin column wheels, this 500-piece limited edition with single-button chronograph and split-seconds capability is worn on a personalized Panerai rubber strap.

PANERAI

LUMINOR 1950 RATTRAPANTE 8 DAYS TITANIO – REF. PAM00530

A 47mm brushed titanium timepiece of distinguished clarity, this hand-wound chronograph boasts a sophisticated 356-component movement entirely executed in-house. Equipped with three barrels and a double column wheel, the P.2006/3 caliber, oscillating at a frequency of 28,800 vph, features an eight-day power reserve and split-seconds chronograph. Engaged by a titanium button at 10:00, the rattrapante hand enables the simultaneous measure of a secondary event to complement the chronograph, whose single-button functionality is performed by a piece at 8:00. A seconds-reset feature permits perfect time-setting synchronization as it zeroes the 9:00 seconds hand and halts the balance while the crown is lifted.

PIAGET

PIAGET POLO FORTYFIVE CHRONOGRAPH – REF. G0A37004

A modern take on the renowned Piaget Polo watch, this handsome and athletic timepiece has a diameter of 45mm, which alludes to the 45-minute duration of a polo match. Piaget shows its' daring side as the case of this model is made of materials that were foreign to the house. Composed of titanium and steel godroons with a black ADLC treatment, the exterior of this watch provides the wearer with maximum comfort. The black dial is complimented with bright white appliqué hour and minute markers as well as luminescent hands which guarantee legibility. Aside from the triple date window at 12 and the second time zone counter at 9, this watch possesses a flyback chronograph complete with a 30-minute counter at 3.

RICHARD MILLE

RM 039 TOURBILLON FLYBACK AVIATION E6B

To explore the themes of aviation and aeronautics in greater depth, Richard Mille has created a caliber designed for flight navigation. This complete and technically sophisticated instrument is an ultra-modern interpretation of an essential device used by every established pilot. The watch displays much of the same information provided by the famous E6B flight computer. This round slide rule, incorporated into the bidirectional rotating bezel, can be used to read off and calculate fuel burn, flight times, ground speed, density altitude or wind correction, as well as the fast conversion of units of measurement. The RM 039 tourbillon caliber is also equipped with a UTC hand, a countdown mode, an oversize date at 12 o'clock and a function selector. The movement drives a flyback chronograph featuring an exclusive design.

RICHARD MILLE

RM 050 FM TOURBILLON SPLIT SECONDS COMPETITION CHRONOGRAPH

The newly designed manual-winding tourbillon movement RMCC1 with hours, minutes, chronograph, split seconds, power reserve, torque indicator and function indicator represents a new progression and development at Richard Mille Watches of the flagship RM008 Tourbillon Split Seconds model. The new RM 050 Chrono Competition Felipe Massa represents a novel movement that was researched to lower the movement weight by 20%, making the total movement weight a mere 9.5 grams. The baseplate and bridges are created from grade-5 titanium, an alloy composed of 90% titanium, 6% aluminum and 4% vanadium often used in the aeronautic, aerospace and automobile industries. Further development of the caliber with its skeletonized baseplate and bridges as well as innovative movement designs meant the creation of more than 450 new parts for the RMCC1 movement. The case is composed of polymer composite injected with carbon nanotubes.

RJ-ROMAIN JEROME

STEAMPUNK RED CHRONO – REF. RJ.T.CH.SP.003.01

Part of the Titanic-DNA collection, the Steampunk Chrono is housed in a Ø 50mm case. The mechanical self-winding movement has been specially developed for RJ-Romain Jerome and is composed of more than 300 elements. Devoid of any dial, this retro-futuristic timepiece features a bezel in oxidized and stabilized steel made with parts of the Titanic. A mobile propeller driven by the oscillating weight enlivens the dial at 6:00, recalling the majestic propellers of the Titanic itself. The minutes and seconds counters echo the aesthetic of ships' manometers and thereby pay tribute to an engine room-inspired, steampunk style.

ROGER DUBUIS

PULSION CHRONOGRAPH – REF. RDDBPU0005

The Pulsion Chronograph explores the stunning visual appeal of transparency with a dial that reveals the technically sophisticated, beautifully finished, automatic-winding RD680 caliber. The monochrome quality of the black DLC titanium case serves to focus the eye upon the intricacies of the movement's functioning as well as its finishing. Granted the Poinçon de Genève, as are all Roger Dubuis watches, the Pulsion Chronograph attracts admiration with a deft dial architecture that renders a complex tableau immediately comprehensible. Worn on a black rubber strap, the Pulsion Chronograph is also available in pink gold and titanium.

TAG HEUER

CARRERA MIKROGIRDER – REF. CAR2C10.FC6323

Conceived, developed and manufactured in-house, the Mikrogirder is the fastest mechanical regulator ever crafted and tested. Accurate to an unprecedented 5/10,000 of a second, the Mikrogirder possesses a completely new regulator system: a coupling beam/girder and excitatory beam system working with a linear oscillator that vibrates at a very small angle, as opposed to a traditional watch, which vibrates at an angle of up to 320°. Ten patents are pending on this breakthrough. The statistics are astounding: 1,000 Hz = 7,200,000 vph. A regular watch functions at 4hz, or only 28,800 vph. TAG Heuer has embedded this mechanical engineering masterpiece in a stunning asymmetric chronograph case design. The unconventional shape and wide opening makes for easier reading, while the dial design allows one to see the beam regulator system at work. This extraordinary model was also the winner of the Aiguille d'Or Grand Prix in 2012, which is the most honorable distinction in the world of haute horology.

TAG HEUER

CARRERA MIKROGRAPH – REF. CAR5040.FC8177

The Carrera Mikrograph 1/100th second chronograph possesses several elements that make it stand out unmistakably from the crowd. It is the first chronograph with a central 100th-of-a-second hand turning one revolution per second and 100 jumps per second (3 times faster than on the historical Heuer model). The dual-chain architecture governs chronograph functions and setting watch time, and the chronograph functions can also be started or stopped independently of watch functions. The chronograph's power reserve is independent of the watch's, due to two separate barrels, but winding the chronograph also winds the watch (which additionally boasts automatic winding). This first 1/100th second wrist-worn stopwatch features an extremely legible chronograph scale on the outside edge of the dial indicating tenths and hundreds of a second. This watch is available in a limited edition of 150 pieces.

VACHERON CONSTANTIN

OVERSEAS CHRONOGRAPH – REF. 49150/B01A-9745

Beating with a frequency of 21,600 vph, the Overseas Chronograph's automatic-winding 1137 caliber provides this timepiece with a precise set of functionalities. At 3:00 and 9:00, respectively, the chronograph's 30-minute and 12-hour counters join the small seconds at 6:00 and large date at 12:00, to complete the backdrop of the frontal view. While the stainless steel case and blue dial give the watch a powerful allure, the adornment of the caseback with the two-masted Overseas ship medallion reminds the wearer of watchmaking's rich nautical heritage.

ZENITH

EL PRIMERO STRATOS FLYBACK STRIKING 10TH – REF. 03.2062.4057/69.M2060

First watch to break the speed of sound in a near space environment with Felix Baumgartner, the El Primero Stratos Flyback Striking 10th is equipped with the world's most accurate automatic chronograph movement: the legendary El Primero. Vibrating at a rate of 36,000 vph, El Primero Calibre 4057 B is framed by a generous 45.5mm case that is water resistant to 100m. A worthy descendant of the Rainbow Flyback model developed for the French Air Force in 1997, this impressive chronograph boasts a flyback function in addition to its capacity to display 1/10ths of a second.

Alarms

Breguet

Marine Royale

The alarm is one of the most useful functions that a wristwatch can offer.

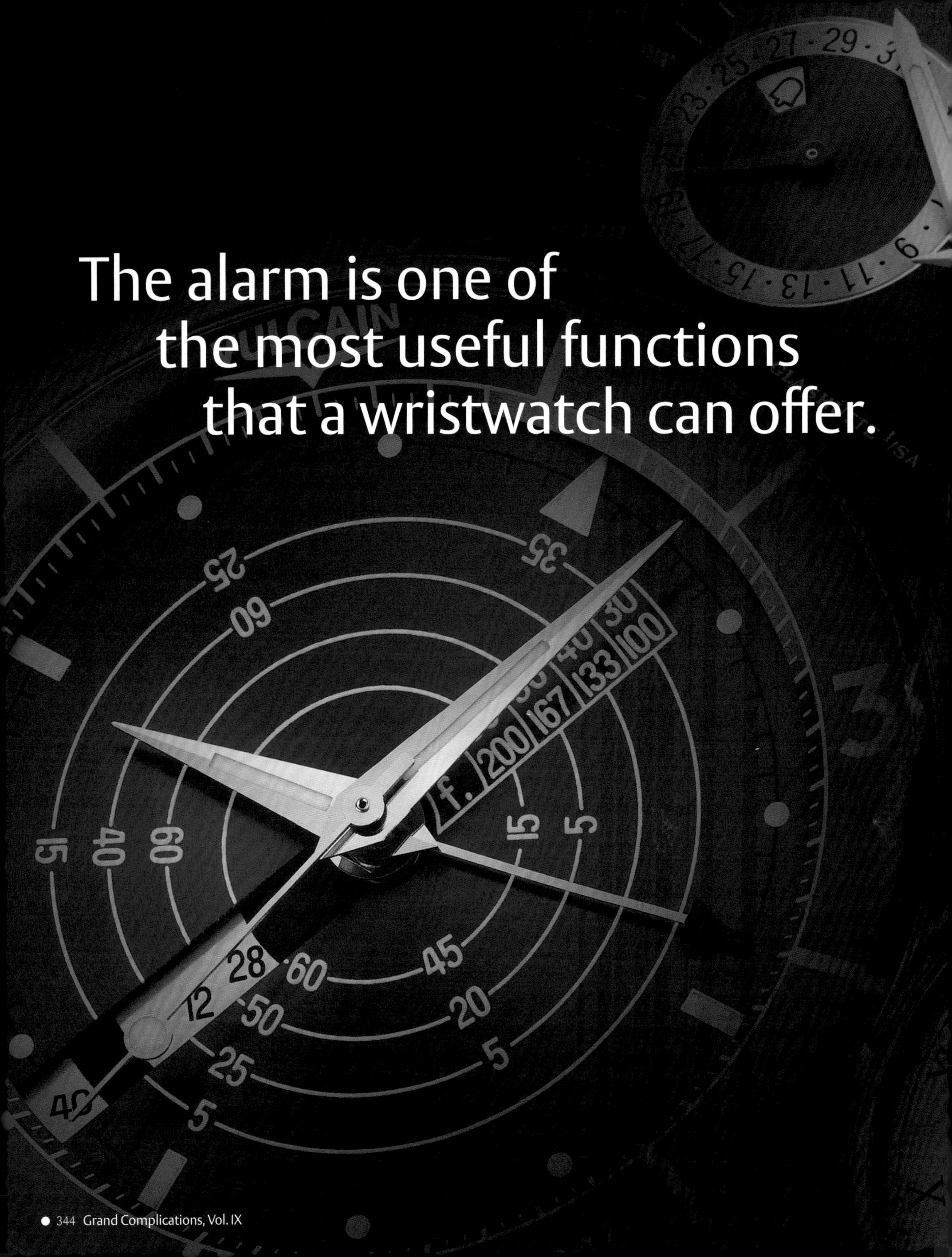

Oris, *Artelier Alarm*

Oris—known for its small alarm clocks in the 1940s—now offers an elegant, classic wristwatch named Artelier Alarm, with a central display separate from the alarm time and a chime specially developed by the brand.

Vulcain, *Golden Heart*

Due to its partially skeletonized dial, Vulcain's Golden Heart allows the wearer to admire the complex architecture of the manufacture's Cricket movement, with its anthracite coating and carefully executed finishings.

Jaeger-LeCoultre, *AMVOX1 R-Alarm*

Born from the partnership between Jaeger-LeCoultre and Aston Martin, the AMVOX1 R-Alarm stands out for its sporty aesthetic that recalls automobile gauges. The alarm is set with a small hand on a minute track at the center of the dial.

Alarms one Can Set in Advance

On a traditional alarm watch, the alarm can only be set 12 hours in advance, i.e. in the evening for the next morning. To increase comfort, several brands have developed mechanisms that can set the alarm 24 hours ahead of time—or even a month in advance. Certain systems also set the alarm to the minute.

Ulysse Nardin, *Sonata Streamline*

On the Sonata Streamline from Ulysse Nardin, the user sets the alarm using an hour/minute counter at 1:30. The 24-hour countdown display at 10:30 allows the alarm to be set the day before with no risk of confusion. The date, alarm and countdown automatically adjust to the selection of a new time zone.

Harry Winston
Project Z6 Black Edition

Harry Winston has put its name to a highly original reinterpretation of the alarm watch in its limited edition Project Z6 Black Edition. The small off-center dial at 4 o'clock allows the wearer to set the hour and minute of the alarm up to 24 hours in advance, using two discs and a day/night indicator. The rectangular gong, cast in one piece, guarantees optimal acoustics.

Glashütte Original, *Senator Diary*

The German brand Glashütte Original, a branch of the Swatch Group, innovated with its presentation of the Senator Diary, the first alarm watch programmable for up to 30 days in advance. The crown at 10 o'clock and the pushbutton at 8 o'clock allow the wearer to determine the date and time of the alarm to within a quarter of an hour.

Musical Alarms

Moving away from more strident sounds, certain brands prefer to charm the ear by equipping their alarm watches with "cathedral" chimes worthy of the most beautiful minute repeaters, or even the kind of precious mechanisms that play entire tunes, as do music boxes. These dulcet tones soften the harsh reality of waking up—or remind the wearer with a few notes of an important meeting during the day.

Breguet, *Réveil Musical*

Masterpiece of haute horology, Breguet's Réveil Musical possesses a patented mechanism inspired by music boxes, with the cylinder replaced by a disc with pins. During the melody, which lasts about 20 seconds, the dial completes a full rotation. The automatic caliber is equipped with a silicon escapement.

Vulcain, *Imperial Gong*

The first wristwatch with an alarm and a tourbillon, Vulcain's Imperial Gong—launched in 2005—also stands out for its cathedral chime, worthy of the most beautiful minute repeaters.

Pierre DeRoche, *TNT RendezVous*

The TNT RendezVous from Pierre DeRoche is not designed to wake a slumberer. It prefers to discreetly signal a moment chosen in advance by a series of double chimes. Its two hammers strike two "cathedral" gongs partially visible through the dial.

Travel Alarms

What is more useful than an alarm while one is traveling? Inspired by this rhetorical question, watchmakers are increasingly combining the alarm function with a second time zone or "world time" display or (see chapter on Multiple time zones and GMTs). The most sophisticated models stand out due to their alarm mechanism that is always linked to the local time.

Breguet, *Le Réveil du Tsar*

The Réveil du Tsar, from Breguet, possesses an alarm mechanism indexed to the local time, with an alarm engagement indicator in a small window at 12 o'clock. The hours/minutes subdial, at 3 o'clock, allows the wearer to set the alarm with to-the-minute precision.

Blancpain, *Léman GMT Alarm*

On the Léman GMT Alarm from Blancpain, the alarm is indexed to the local time display, indicated by the large hour and minute hands. Another practical innovation: the automatic-winding mechanism allows the wearer to charge the power reserve and the alarm simultaneously.

Zenith, *Class Traveller Multicity Alarm*

The Class Traveller Multicity Alarm from Zenith associates the famous automatic-winding El Primero chronograph movement with an alarm and a world time display. It is equipped with two barrels: the first for the alarm, the second for the model's other functions.

Panerai
Radiomir GMT Alarm

The Radiomir GMT Alarm from Panerai combines a digital display of a second time zone at 6 o'clock with an alarm indicated by a central hand. The two functions are controlled by a crown located on the right side of the case.

Yeslam, *Ti2 Réveil*

Equipped with a movement from La Joux-Perret, the Ti2 Réveil from Yeslam contains within its square titanium case an alarm with a central hand, an aperture for the date and a 24-hour time zone with a digital display.

Vulcain, *Revolution GMT Automatic*

The Revolution GMT Automatic from Vulcain associates an automatic-winding alarm movement with a world time display. The ring indicating the 24 reference cities for the 24 time zones can be set using a crown at 4 o'clock.

Jaeger-LeCoultre, *Master Compressor Extreme W-Alarm*

The Master Compressor Extreme W-Alarm from Jaeger-LeCoultre, with a world time function, presents several patented innovations in its alarm mechanism, which is equipped with a gong wrapped around the movement. The time of the alarm can be precisely selected using a double disc with hours and minutes in a window at 9 o'clock.

Underwater Sounds

An ever-increasing number of diving watches are equipped with an alarm, and more and more alarm watches are getting their sea legs. This might seem like an odd alliance, as the ocean depths are not ideal for spending the night, or even for a small nap. But here the alarm is used to remind the wearer of a precise moment, notably the end of the dive time—and it doesn't hurt that sound carries much better underwater than in open air.

Jaeger-LeCoultre
Memovox Tribute to Deep Sea

In homage to Memovox Deep Sea, the first Jaeger-LeCoultre diving watch, released in 1959, the Memovox Tribute to Deep Sea is powered by the automatic Memovox 956 caliber, with hours, minutes, seconds and alarm. The caseback reprises the timepiece's original design of a frogman surrounded by bubbles.

Vulcain, *Nautical Heritage*

In a rerelease of a famous 1961 model, the Nautical Heritage from Vulcain—water resistant to 300m—stands out for its alarm that is audible underwater, its central rotating dial displaying decompression stages and its triple-backed case, which serves as a resonance chamber for the chime.

Vulcain

Diver X-Treme Automatic

Vulcain has developed several watches that combine an alarm with underwater capability, such as the Diver X-Treme Automatic, with a case that is water resistant to 100m, audible underwater strike, rotating 60-minute bezel to calculate the dive time and an ultra-readable dial highlighted with luminescent indexes.

Breguet, *Marine Royale*

Breguet has released its sporty Marine Royale collection in an alarm model water resistant to 300m, with a modern aesthetic that combines rubber and rose gold.

Jaeger-LeCoultre, *Master Compressor Diving Alarm Navy SEALs*

Jaeger-LeCoultre celebrated its partnership with the elite combat swimmers through the 2009 launch of the Master Compressor Diving Alarm Navy SEALs, a limited edition with a grade 5 titanium case, a ceramic bezel and water resistance to 300m.

BREGUET

CLASSIQUE LA MUSICALE – REF. 7800BA.11.9YV

A masterful fusion of innovation, sublime craftsmanship, and design ingenuity, this self-winding 48mm timepiece in 18K yellow gold conducts a brilliant visual accompaniment to its pioneering musical score. Equipped with the first ever magnetic strike governor, a patented system of isolated magnets used to eliminate background noise, the melodic tune, pre-set to a desired time or activated via a pushbutton at 10:00, is joined by the superb and complete rotation of the hand-guilloché dial. Powered by the 55-jewel 0900 Caliber with silicon escape wheel and Breguet balance and spring, this horological work of art is finished with a caseband intricately engraved with a musical stave.

BREGUET

MARINE ROYALE – REF. 5847BR.Z2.5ZV

Capitalizing on water's superior sound conduction, carrying tones approximately four times faster than above the surface, this 45mm 18K rose-gold timepiece brings sonorous enchantment to the world of diver's watches. An authentic diving instrument first and foremost, the self-winding timepiece boasts a unidirectional rotating bezel, as well as luminously enhanced 18K gold hands, optimized for legibility in murky, submerged conditions. Set at 4:00, the alarm's status is discreetly clarified via an aperture at 12:00 on the black rhodium-finished 18K gold dial with rose-engine engraved wave pattern. Armed with a 36-hour power reserve, the 519R caliber boasts an 18K rose-gold rotor.

HARRY WINSTON

PROJECT Z6 BLACK EDITION – REF. 400/MMAC44ZKC.K2

Equal parts striking and practical, the Project Z6 Black Edition is a flawlessly sporty and striking timepiece. It is crafted in Zalium™ with a black DLC treatment, a material exclusive to Harry Winston. This non-allergenic, corrosion-resistant alloy is exceptionally durable and light as it is harder than titanium. This treatment not only adds luster, but rigidity too. The movement of the Project Z6 is as stunning as its case. The alarm—designed to be seen and heard—is fixed by a visible hammer that is attached to the case to showcase the alarm as it rings. The alarm is set with the crown and activated by using a repeater-style trigger positioned on the case at 4:00. When the manual movement is fully wound, the alarm will ring for a complete 20 seconds, underlining the practicality of this cutting-edge timepiece.

ZENITH

PILOT DOUBLEMATIC – REF. 03.2400.4046/21.C721

The Pilot Doublematic is a open invitation to travel. This chronograph exercises infallible control over time and features an alarm function controlled by the pushbutton and the crown at 8:00. The Pilot Doublematic also defies time differences thanks to its two discs simultaneously displaying the time in different parts of the world. Beating at 36,000 vph and endowed with a 50-hour power reserve, the mechanical automatic-winding El Primero Calibre 4046 may be admired through an openworked oscillating weight that is itself visible through the sapphire caseback.

Multiple Time Zones and GMTs

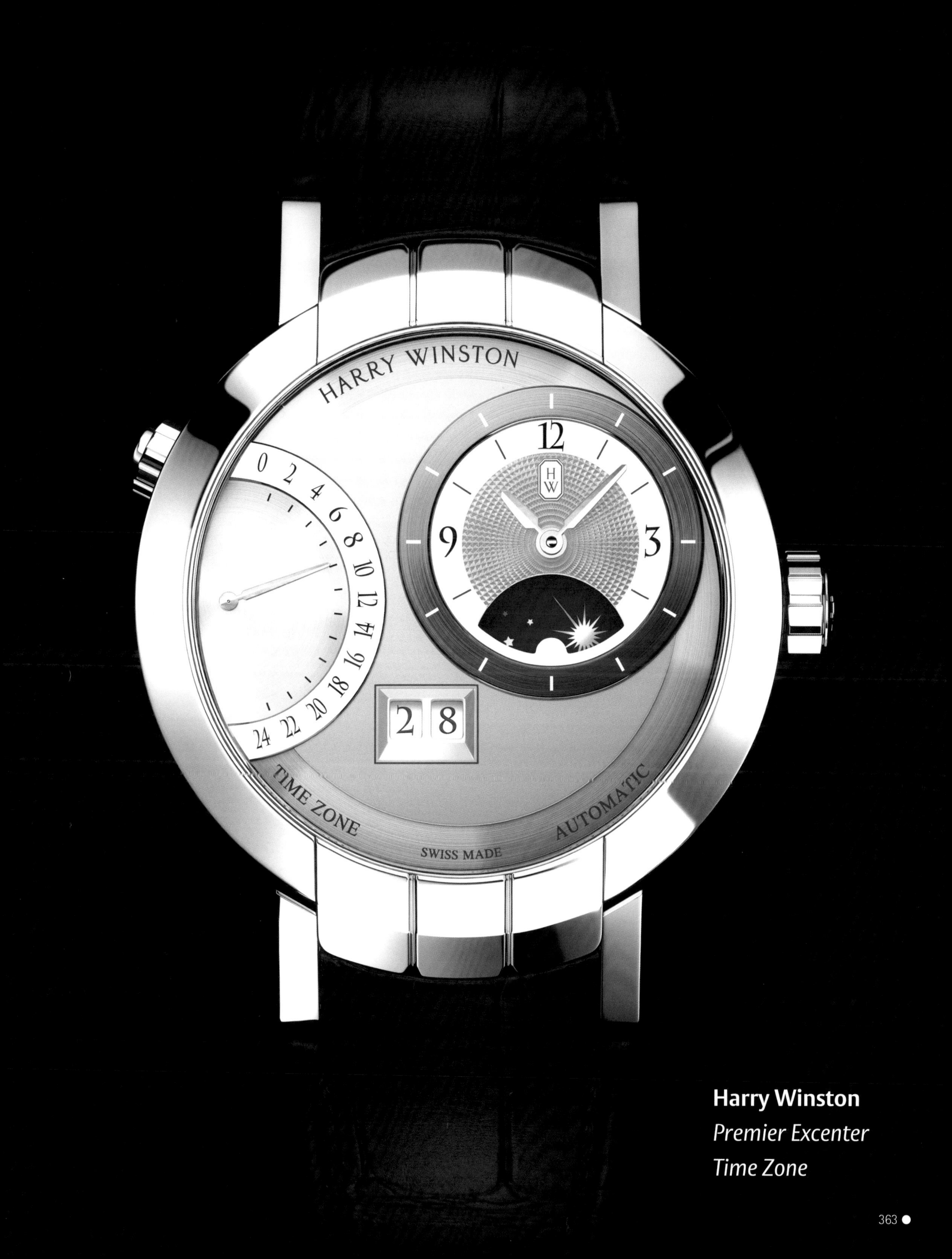

Harry Winston
Premier Excenter
Time Zone

Fawaz Gruosi

Founder and Creative Director of de Grisogono

"I'm still ready to take risks, but they are more carefully considered.""

20 YEARS OF CREATIVITY

Very soon after its founding in 1993, de Grisogono brought a fresh and innovative breath of fresh air to the jewelry world. Whether through a gem-setting technique that rendered the underlying claws invisible, the unique use of a distinctive palette of colors or its range of unique designs, the brand has striven to create the unexpected in both its watches and its jewelry. Nonetheless, the true driving force behind this creativity lies in Fawaz Gruosi's ability to buck trends. Following his own instincts and refusing to get stuck in a commercial-thinking rut, he has taken risks that were unprecedented at the time, such as the use of black diamonds or pink gold at a time when it was neglected by other brands. It is through these daring—and winning—gambles that the brand has managed to capture public attention and express its creativity. This taste for risk-taking has thus become an authentic strength that has enabled the company to define its own style. Gruosi recalls: "We have created exceptional objects, akin to works of art, for which the cost was high in this sector of the economy. By refusing any kind of minimalism and by paying attention to every detail, we began creating haute couture pieces in the field of jewelry and then in watchmaking."

Watchmaking is an adventure that began in 2000 for de Grisogono, and the brand immediately made an indelible and original mark on this often-conservative field. For Gruosi, this breakthrough into the world of horology opened up whole new territories to explore, with creativity was given a prominent role. This approach is clearly apparent in the design of the collections, as well as in the launch of extremely complex mechanical watches such as the Occhio Ripetizione Minuti, Meccanico DG and Otturatore. For women, models such as Instrumentino, Piccolina, Tondo by Night and Tondo Tourbillon have had a particularly notable impact. If Gruosi were to choose just three watches from these first 13 years in watchmaking, he would single out the Instrumento No Uno, launched in 2000, which, as he explains, "will always be de Grisogono's first watch." When it comes to the Instrumentino, he describes it "as the first watch ladies' watch, released in 2002, which thereby holds a place of its own." Finally, the Doppio model, introduced in 2003, is viewed by its designer "as the first de Grisogono watch with horological complications"—and a model that paved the way for a prestigious line of successors.

GROWTH AND CHANGE

In 20 years, de Grisogono has expanded from its small early structure with around ten employees to a company with a 135-strong staff. This growth has naturally entailed significant changes, the most notable of which came in 2012 with the entry of new shareholders into the company's capital. "Despite our growth, de Grisogono has maintained an almost family-style team spirit. Decisions have been made gradually, once our shareholders had taken the time required to understand how our company worked." The brand has thus maintained its taste for challenges and continues to assert its innovative nature. Nonetheless, Gruosi admits to a slight shift in his personal approach: "With the development of de Grisogono, I have acquired a sense of being entrusted with new responsibilities.

MECCANICO DG

I am the guarantor for my employees and each decision may have repercussions on 135 people who are married and have families. I'm still ready to take risks, but they are more carefully considered."

The future development of de Grisogono is already clearly planned. In light of the current economic context, the company has chosen to elaborate a well-defined strategy. In respect to the ongoing evolution of the firm, Gruosi sums it up in these terms: "In 2012, we took the time to plan our development, to organize our infrastructure and to make key decisions in order to develop our strong points. For 2013, we are going to put these modifications in place and to implement new methods. Finally, in 2014, people will be able to discover the results of the changes that have been made." de Grisogono is thus all geared up for a new evolution. Nonetheless, far from resting on its laurels, the brand has already made the most of its inventive spirit in presenting a new creation unveiled in February 2013: the Sugar collection.

SWEET TREATS ON WOMEN'S WRISTS

With the Sugar collection, de Grisogono has created a resolutely feminine and precious watch with a classic yet modern style. Inspired by the inimitable elegance of a woman pouring sugar into her coffee, these square timepieces showcase the brand's gem-setting talent, since they are entirely adorned in precious stones on the bezel, the case back and sides, and the bracelet buckle. The various sizes of diamonds used vividly evoke a sugar cube. Nonetheless, this model's most significant technical accomplishment lies between the bezel and the galuchat strap. A mobile gem-setting technique enables the addition of diamonds that—in step with the wearer's every move—create the illusion of a shower of stars flowing over the wrist. The dial of this model comes in two versions: mother-of-pearl or entirely paved with precious stones. Famed for its ability to play with colors, de Grisogono makes no exception to its rule with Sugar, offering models set with blue sapphires, orange sapphires, white diamonds or emeralds. Each of these variations is available with a pink- or white-gold case measuring 36x36mm. With Sugar, de Grisogono continues to explore the territory charted by the Tondo collection and offers feminine timepieces equally well suited to glamorous society events as to relaxed contemporary evenings.

While Sugar is in the spotlight in early 2013, the year's BaselWorld Fair witnessed the arrival of a revamped Instrumento No Uno with a large date and a new dial. For fall of 2013, Gruosi announces the launch of a new ladies' watch featuring an original design that is bound to attract notice. Finally, the year 2014 is already set to produce some striking new watch models, not to mention several other surprises that will of course accompany the upcoming de Grisogono launches.

SUGAR

Symbols of the conquest of the air, and of fast-paced globalization,

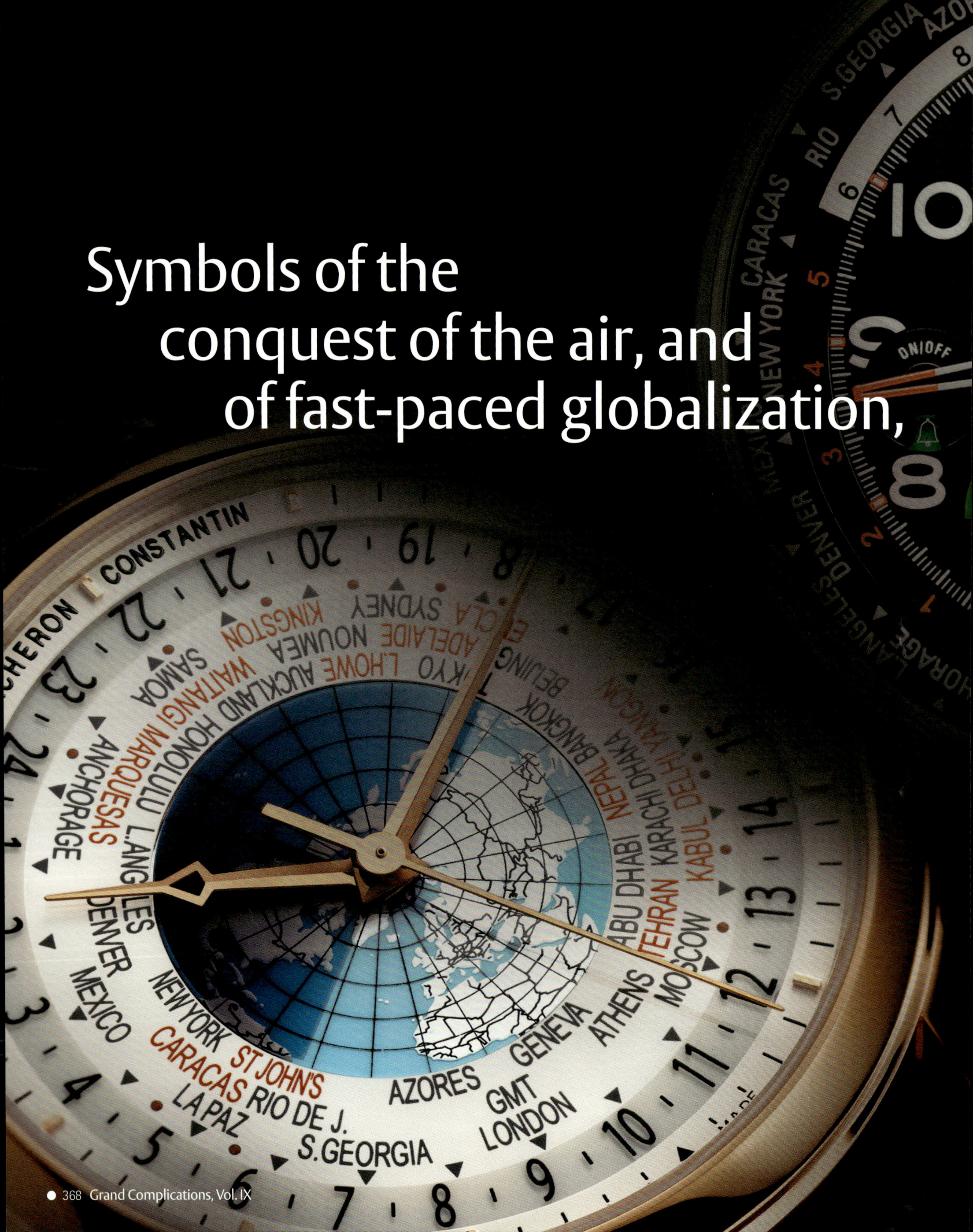

watches that display several time zones are now meeting with unprecedented success. Of all the horological functions, this is without a doubt the one that presents the most varied modes of display. From a simple watch with an extra hand to the famous "world time" functions that indicate the hour across the globe, ingenious watchmakers compete to propose the most practical or most original systems. We are also seeing an increasing number of models that take into account half- or quarter-time zones—as well as timepieces equipped with miniature three-dimensional planets.

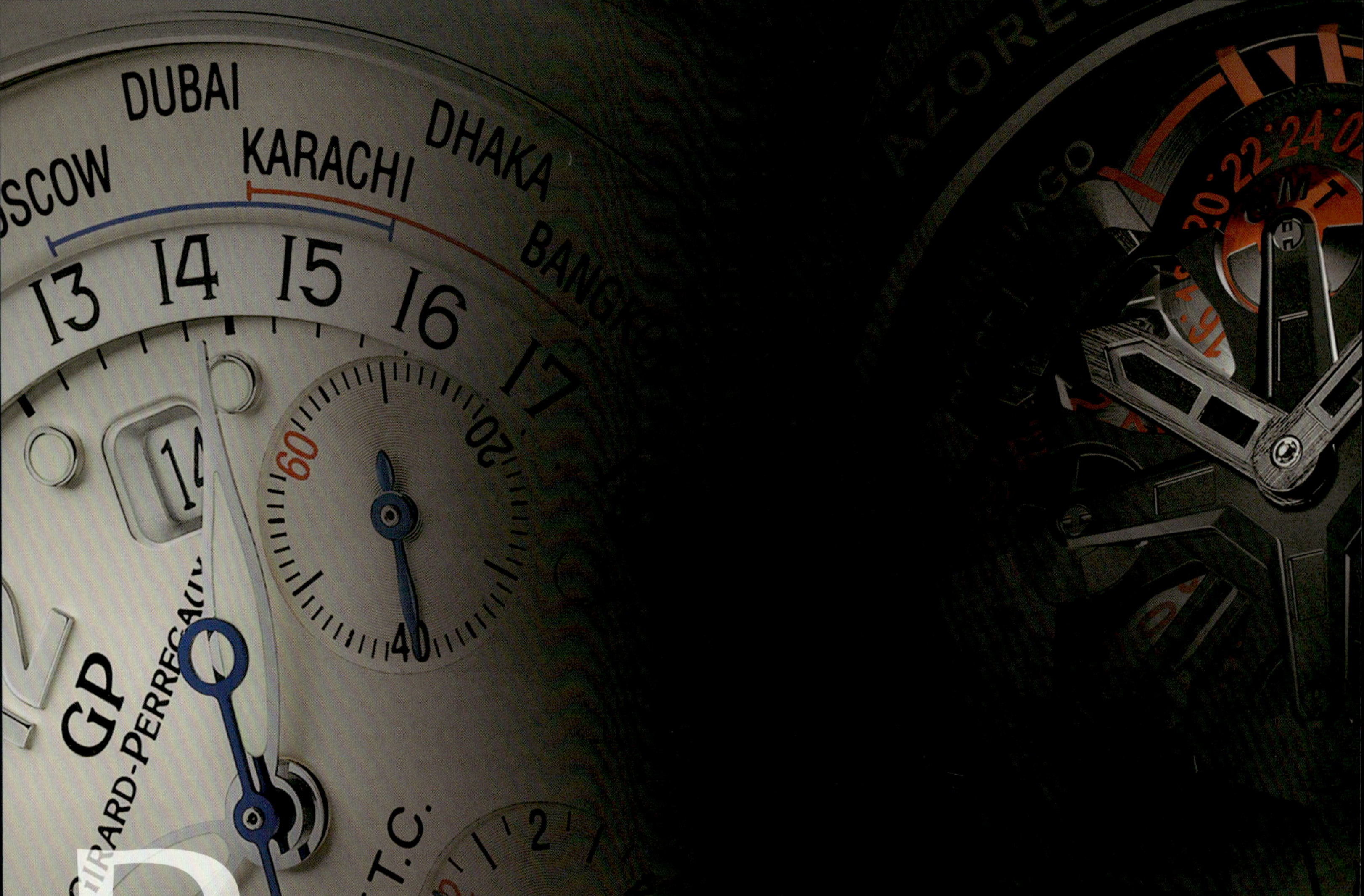

Before the division of the globe into 24 time zones, each region lived at its own pace, and travelers had to adjust their watches constantly as they moved from one place to another. There were a multitude of local times, based, more or less, on longitude: more than 70 in North America and around 30 in Europe. The increase in long-distance travel, and especially the birth of the railroad, generated a need to create a unified system of time. In 1847, the British Railway Clearing House recommended that all companies adopt the time at the Greenwich Observatory (Greenwich Mean Time, or GMT), which already served as a reference for navigators. In 1884, at the International Meridian Conference in Washington, DC, the decision was taken to divide the planet into 24 time zones, each one covering 15° of longitude (one hour), and to assign longitude 0 to the Greenwich meridian. This system was rapidly adopted by most of the countries in the world—except France, which did not come around until 1911.

The first dual time zone watches, sometimes called "captain's watches," were equipped with two movements, more or less coordinated, that displayed the time zones on two different dials. Later, watchmakers perfected this system by using just one movement, equipped with a mechanism uncoupling the additional hour hand.

Though watches with a multitude of local times already existed as early as the 17th century, it was not until the 1930s that a Genevan watchmaker, Louis Cottier, invented the ingenious "world time" system, which allows the wearer to read the time in all time zones simultaneously.

GMT AND UTC

Greenwich Mean Time corresponds to the mean solar time at the meridian crossing the Greenwich Royal Observatory, near London. It was adopted at the end of the 18th century as a reference time zone for all others on the planet, and is based on the rotation of the Earth. It was replaced in 1972 by "Universal Coordinated Time," better known by its acronym UTC, which is based on International Atomic Time (with the acronym TAI, for its French name). Watchmakers commonly use the term GMT to designate a watch indicating a second time zone, even if that time zone is not Greenwich.

TECHNIQUE

HOW DOES IT WORK?

Multiple time zone watches offer such a variety of technical approaches and methods of use, that it is impossible to describe in detail all the particularities of their functioning.

Generally, they possess an additional display for the hour that is powered by the main system, but can be temporarily disengaged in order to set it to local time—without affecting the accuracy of the minutes and seconds. A pushbutton or the crown serves as a tool for setting the second time zone.

In order not to awaken a friend or a business partner who might be sound asleep, travelers often need to know if it is day or night in their home region. This is why some watches display home time on a 24-hour scale. Others use a 12-hour display, but supplement this with a day/night indication (a small aperture or subdial that changes from white to black according to the time of day), a small additional 24-hour scale or an AM/PM display.

Many watches boast another very useful function when one is traveling: automatic linkage of the date to local time.

WORLD TIME

To continuously display the time in all corners of the globe, world-time watches possess—in addition to the central hands—two mobile discs, one with 24 gradations, the other bearing the names of the reference cities. The 24-hour disc turns counterclockwise.

Double Pointer-Type Systems

The simplest system for simultaneously displaying two time zones is to add an hour hand with a distinctive design, either to the center of the dial, or—more rarely—on a small auxiliary dial. The second time zone might also distinguish itself with a retrograde or jumping display. The most practical models equip their home time zone with a 24-hour scale or a day/night indicator. Luxury watchmakers also take great care to make the systems for setting the time as easy to use as possible.

Omega, *Seamaster Aqua Terra GMT*

Equipped with an exclusive co-axial escapement movement, the Omega Seamaster Aqua Terra GMT displays a second time zone on a 24-hour scale using a red-tipped hand. The vertical relief on the dial evokes the teak wood interiors of luxury boats.

On the Fuso Quadrato from de GRISOGONO, the hand for the additional time zone is hidden beneath a dial composed of 12 titanium shutters that opens on demand using a slide at 9 o'clock.

Jaquet Droz, *Les Longitudes*

Les Longitudes from Jaquet Droz draws inspiration from regulator-style dials, with a Roman numeral subdial at 3 o'clock for the first, 12-hour time zone, a 24-hour display with Arabic numerals at 9 o'clock for the additional time zone, and a central minute hand.

Viator, *Heritage Watch Manufactory*

On the Viator, from Heritage Watch Manufactory, equipped with a manufacture-made movement that boasts three patented mechanisms, the second time zone appears in an instantaneous jumping display with a quick-correction system.

DeWitt, *Academia Double Fuseau GMT II Poetic*

The Academia Double Fuseau GMT II Poetic from DeWitt offers an original GMT complication. The wearer sets the second time zone, indicated by a red central hand, by pinpointing its position on the tiny globe at 12 o'clock.

Rolex, *Oyster Perpetual GMT-Master*

A classic example of the dual time zone watch, the Rolex Oyster Perpetual GMT-Master—launched in 1955—resolves the problem of daytime or nighttime with a supplementary time zone that covers all 24 hours.

Harry Winston

Premier Excenter Time Zone

The Premier Excenter Time Zone model, from Harry Winston, associates an off-center hours/minutes display for the first time zone (with day/night indicator) with a retrograde second time zone on a 24-hour scale on the left side of the dial.

F.P. Journe, *Octa UTC*

The Octa UTC from F.P. Journe uses as a reference not Greenwich Mean Time, but UTC (Coordinated Universal Time), which replaced it in 1972. The small mobile planet at 7 o'clock takes account of summer (DST) and winter times.

Superimposable Hands

On most travel watches, the additional time zone hand is always visible, even when the wearer has no use for it. Some watchmakers have improved upon this "defect" by devising hour hands that can be superimposed, or placed one on top of the other, and function in a perfectly synchronized manner. This is a means of offering two watches in one.

H. Moser & Cie., *Meridian Dual Time*

On the automatic Meridian Dual Time from H. Moser & Cie., the two superimposable hands are set via the crown. The digit 12 or 24 in the aperture at 12 o'clock indicates whether it is day or night (AM or PM) at the traveler's home.

Patek Philippe,
Aquanaut Travel Time Reference 5164

On the Aquanaut Travel Time from Patek Philippe, the pushbuttons allow the wearer to bring the local hour hand forward or back in increments of one hour. When the dual time zone function is not in use, the two hour hands appear as one.

Montblanc
Grand Chronographe Régulateur

The Grand Chronographe Régulateur from Montblanc boasts two superimposable hour hands in a subdial at 12 o'clock, as well as a day/night indicator linked to home time.

Jaeger-LeCoultre, *Master Hometime*

The Master Hometime from Jaeger-LeCoultre proposes a system of dual crown-activated superimposable hour hands. The skeletonized hand on a 24-hour scale maintains home time. The date is linked to the local time.

Half- or Quarter-Time Zones

On the vast majority of travel watches, the second time zone is adjustable in increments of one hour. But there are countries in which the deviation from UTC adopts finer divisions of half an hour or even a quarter of an hour. This is notably the case with Iran (UTC + 3h30), India (UTC + 5h30), Nepal (UTC + 5h45) and the middle of Australia (UTC + 9h30). A few recent watches take these particularities into account.

Franck Muller, *Master Banker*

The Master Banker from Franck Muller presents, in addition to a central hour, two additional time zones adjustable to the minute, each one accompanied by a day/night indicator.

Blancpain

Villeret 8 Day Half-Timezone

The Villeret 8 Day Half-Timezone from Blancpain allows the wearer to set the second time zone (which includes hours and minutes) in increments of half an hour using a crown equipped with an integrated pushbutton.

Glashütte Original

Grande Cosmopolite Tourbillon

The Grande Cosmopolite Tourbillon from Glashütte Original can display two hour/minute time zones from among a selection of 37 locations (including those that vary by 30 or 45 minutes), with a distinction between summer and winter times. Thanks to its perpetual calendar, it also takes into account the exact monthly and leap-year cycle for these calculations.

Parmigiani, *Tonda Hémisphères*

Parmigiani's Tonda Hémisphères, with a double hour/minute display and two day/night indicators, allows the wearer to set the second time zone to the nearest minute.

Double Hour/Minute Displays

Though the minutes generally remain the same from one time zone to the next, certain high-end watch brands prefer to include a complete display of the time in each region, with a second set of both hour and minute hands located in a small auxiliary dial. Another possibility involves indicating the two time zones on two separate dials, in watches featuring a resolutely distinctive design.

de GRISOGONO, *Meccanico dG*

Newly available in black titanium, de GRISOGONO's astonishing Meccanico dG combines an analog display for the first time zone and a digital display, driven by 23 cams, for the second time zone.

Breguet, *7067 Tradition GMT*

Equipped with an openworked architecture that reveals the principal components of the movement, the 7067 Tradition GMT from Breguet displays local time at 12 o'clock. A guilloché subdial at 8 o'clock displays home time, complete with day/night indicator.

Audemars Piguet
Jules Audemars Dual Time

The epitome of elegance and refinement, Audemars Piguet's automatic-winding Jules Audemars Dual Time displays its second time zone on a subdial at 6 o'clock, overlapping with a day/night indicator.

Jaeger-LeCoultre, *Reverso Grande GMT*

The Reverso Grande GMT from Jaeger-LeCoultre makes the most of its famous reversible dial by proposing two time zones on two dials that lie back to back. The dial on the reverse side includes an indicator for the number of hours' time difference between local time and Greenwich Mean Time or another reference point.

Piaget, *Altiplano Double Jeu*

Piaget's Altiplano Double Jeu possesses two superimposed cases, each containing an ultra-flat mechanical movement. The dial of the lower case displays a second time zone on a 24-hour scale.

Icelink, *Ambassador*

The brand Icelink boldly opts for an extra-extra-large size in its Ambassador line—the better to fit six quartz movements that display the time in six cities around the world.

Systems with Discs or Rotating Bezels

Watches with multiple time zones might operate using systems based on discs or rotating bezels—either in lieu of a pointer-type system or in conjunction with it. Watchmakers generally take this opportunity to introduce complementary indications, such as the names or abbreviations for certain cities, representing precise time zones.

Breguet
Classique Hora Mundi

The Classique Hora Mundi from Breguet displays two preselected time zones, passing from one to the other with a simple press on a pushbutton. The system is completed by a day/night indicator and a date that is linked to the displayed time.

Jaquet Droz, *Les Douze Villes*

The Les Douze Villes watch from Jaquet Droz unites a disc bearing the names of twelve cities with a jumping hour display. A little pressure on the pushbutton displays the name of the desired city and its corresponding time, with minutes indicated by the large central hand.

JeanRichard, *2Timezones Zirconium*

On the 2Timezones Zirconium by JeanRichard, the first time zone is displayed traditionally, in the center, while the second appears in a window at 12 o'clock, accompanied by an AM/PM indication. An aperture at 6 o'clock reveals the 24 cities representing the world's time zones.

Cartier, *Calibre de Cartier Multiple Time Zone*

On the Calibre de Cartier Multiple Time Zone, the user can select the local time using a disc that is visible on the case side—that takes summer and winter times into account. An aperture at 6 o'clock indicates the number of hours between home time and local time.

Hublot, *King Power Unico GMT*

Hublot has equipped its King Power Unico GMT chronograph with four rotating aluminum discs, indicating as many time zones (on a 24-hour scale), from among a selection of 40 cities. The system is governed by a pushbutton situated at 2 o'clock on the case middle.

Vogard, *Datezoner*

The Datezoner from Vogard stands out for its patented system that allows the wearer to change the time zone and the date simultaneously, via the bezel. One needs only to pull the security lever and rotate the bezel to select the desired city.

Corum

Admiral's Cup Challenge 48 Day & Night

The Admiral's Cup Challenge 48 Day & Night from Corum is adorned with a map of the world that bears a transparent rotating disc. This system can perform two functions: it can display a second time using a small white indicator, or it can indicate which side of the Earth is illuminated by the sun and which is in darkness.

Bovet Fleurier

Rising Star Triple Time Zone Tourbillon with Reversed Hand-Fitting

The Rising Star Triple Time Zone Tourbillon with Reversed Hand-Fitting model from Bovet Fleurier displays two additional hour/minute time zones, with indications of the names of the reference cities and day/night indicators.

A. Lange & Söhne

Lange 1 Time Zone

A. Lange & Söhne's Lange 1 Time Zone possesses two auxiliary hour/minute dials with day/night indicators. The wearer adjusts the second time zone with a pushbutton that turns the rotating city ring.

Three-Dimensional Planets

After the appearance of three-dimensional moons (see chapter on Moonphases), a new type of travel watch emerged, one that displayed small terrestrial globes rotating on the dial. These innovative, spectacular creations are most often produced by young, pioneering brands or independent watchmakers.

Greubel Forsey, *GMT*

The titanium globe adorning Greubel Forsey's GMT allows the wearer to take in at a glance all the Earth's time zones—complemented by a day/night indicator. For a more precise reading, the watch also displays a second time zone at 10 o'clock and a world time indication on the caseback.

Hysek, *Colosso*

On the Colosso from Hysek, the terrestrial sphere, rotating over 24 hours, can be linked to either local time (on the central hands) or to a second time zone. The latter is indicated by a double linear retrograde display, with an aperture that specifies the reference city.

Magellan Watch

Magellan 1521 NH

Using a third hand tipped with a small solar disc, the Magellan 1521 NH displays either the position of the sun in relation to the Earth (longitude where it is noon), local time on a 24-hour scale or a second time zone, also on a 24-hour scale.

VicenTerra, *GMT-3*

On the VicenTerra GMT-3, the small rotating sphere placed at 5 o'clock shows the longitude where the sun is at its height. This unusual system is completed by a second time zone at 7 o'clock and a day/night indication (sun and moon) at 12 o'clock.

World Times

Queen of travel watches, the world-time complication allows the wearer to see what time it is in all 24 time zones. To this end, they are usually equipped with a mobile ring that bears the names of 24 reference cities. A few watchmakers have chosen to reinterpret this classic model by varying the configuration of the dials, taking partial time zones into account or displaying specific time slots, such as the opening of the world's stock markets. The difference also finds scope for expression in the adjusting systems, whose practicality varies with the model.

Zenith, *Pilot Doublematic*

Equipped with a black and white "aviator" look, the Pilot Doublematic from Zenith combines world-time chronograph and alarm functions—all powered by an automatic El Primero movement housed in the steel or rose-gold case.

Vacheron Constantin

Patrimony Traditionnelle World Time

The Patrimony Traditionnelle World Time by Vacheron Constantin simultaneously displays no fewer than 37 time zones—including those that differ from their neighbors by a half-hour or a quarter-hour. This exclusive piece is patented.

Breitling, *Transocean Chronograph Unitime*

On the Transocean Chronograph Unitime from Breitling, driven by an in-house made movement, the wearer need only turn the crown to the front or back to correct all indications with one movement—with an automatic change in the local date.

Jean-Mairet & Gillman, *Sport*

In designing the Sport model, Jean-Mairet & Gillman placed the world-time display on a subdial at 9 o'clock, completed by a date window at 3 o'clock—all controlled by a multifunction crown.

Patek Philippe

World Time Ref. 5131

Heir to a long tradition, the World Time from Patek Philippe is distinguished by its dial in cloisonné enamel. To change the main time zone, the wearer uses the sole pushbutton, at 10 o'clock—the mechanism then simultaneously adjusts the city disc, the 24-hour disc and the hour hand.

IWC, *Pilot's Watch Worldtimer*

The Pilot's Watch Worldtimer from IWC is distinguished by its system of making adjustments via the crown—which affects the 24-hour ring as well as the reference time zone displayed by central hands.

Girard-Perregaux

ww.tc Financial

Dedicated to financial whizzes and nervous stock dabblers, the ww.tc Financial from Girard-Perregaux displays—in addition to the 24 time zones—the opening and closing times of the four main stock markets: Wall Street, London, Tokyo and Hong Kong.

Audemars Piguet

Jules Audemars Metropolis

The Jules Audemars Metropolis by Audemars Piguet combines a world-time mechanism with a perpetual calendar. The two rotating discs of the world-time display occupy a subdial at 6 o'clock.

Frédérique Constant

Worldtimer

Offered at a particularly attractive price point for a world-time model, the limited edition Worldtimer from Frédérique Constant houses a manufacture-made movement, automatic movement beneath a dial adorned with a map of the world and a date hand.

Antoine Preziuso

Transworld

Transworld, from Antoine Preziuso, offers a very attractive reinterpretation of world time, with a mobile world map in the center of the dial, depicting the Earth as seen from the South Pole and making a complete rotation over 24 hours. The display is as original as it is poetic.

A. LANGE & SÖHNE

LANGE 1 TIME ZONE LUMINOUS – REF. 116.039

Operating on the solid silver dial of the 18K white-gold Lange 1 Time Zone, the luminous rhodiumed-gold hands and appliqués on this timepiece guide their owner even in the darkness of night. Conveniently adjusted via a pushbutton, the peripherally rotating cities ring offers a cosmopolitan and clear view of the applicable time zone's chosen reference location. Furthermore, this manual-winding timepiece is armed with a 72-hour power reserve, and boasts an ingenious synchronization mechanism that permits the seamless swapping of the indicated time between the two subdials, both of which possess an independent day/night indicator.

A. LANGE & SÖHNE

SAXONIA DUAL TIME – REF. 385.026

Powered by a completely redesigned self-winding caliber L086.2 with a 72-hour power reserve, the Saxonia Dual Time clearly indicates two time zones on the dial. The solid gold hour hand that indicates local time can be moved forward or backward in one-hour increments by two pushbuttons located at 8:00 and 10:00. A supplementary hour hand in blued steel permanently keeps time at the place of origin. The pink- or white-gold 40mm case is fitted with an antireflective sapphire crystal on the front and back. The timepiece is presented on a hand-stitched crocodile strap with a solid gold buckle.

AUDEMARS PIGUET

ROYAL OAK "CONCEPT GMT" TOURBILLON – REF. 26560010.00.D002CA.01

Clad in state-of-the-art grade-5 titanium, this timepiece leans on impressive innovation to adorn the wrist with elegant simplicity. Driven by the remarkably finished manual-winding 2913 caliber, the watch features an original function indicator at 6:00 ("H" for hours and minutes, "R" for winding, and "N" for neutral), controlled by a three-position crown. Opposing the tourbillon at 9:00, the ingenious secondary time zone at 3:00 is comprised of two overlapping discs. While the numbered disc, which indicates 1 through 12, makes two revolutions per day, the disc beneath it, visible via a transparency effect, indicates, as a white or black backdrop, the occurrence of day or night.

de GRISOGONO

MECCANICO S01

de GRISOGONO has created a world's first with the Meccanico dG's mechanical digital display. Powered by a movement comprising 651 elements, the Meccanico dG features two time zones: an analog display of hours and minutes in the upper part and a mechanical digital display of the second time zone in the lower part. The workings of this breakthrough have been patented by de GRISOGONO. A curved rectangular case of 18K pink gold houses the exclusive manual-winding dG 042, a de GRISOGONO caliber with a 35-hour power reserve. Water resistant to 30m, the case is fitted with a transparent black dial with 18K pink-gold dauphine hands, and mounted on a black natural rubber strap with an 18K pink-gold butterfly buckle.

de GRISOGONO

FUSO QUADRATO NO. 3 (CLOSED)

This metamorphosing timepiece draws its power from the DF 21-90 self-winding caliber, oscillating at a frequency of 28,800 vph. The gently rounded square case in 18K rose gold sets the frame for a decisively contemporary black dial upon which dauphine hour and minute hands star in their golden character against the layered face and its titanium shutter center. The modern Arabic numeral layout leaves the watch uncluttered, with only the 3, 6 and 9 acting as indexes for the hands' revolution.

de GRISOGONO

FUSO QUADRATO NO. 3 (OPEN)

Upon activation via the sliding lever at 9:00, the Fuso Quadrato opens its 12 titanium shutters and reveals the GMT 24-hour time zone beneath the initial presentation. The black pointer joins the two 18K rose-gold hands at the center of the watch, indicating the time at a chosen location. This lower face is intricately decorated with a rose-gold Clous de Paris motif that vibrantly plays off the sharp black qualities of the periphery of the face. A picture of ingenuity, contrast and intelligent architecture, the Fuso Quadrato offers a veritable insight into the genius of de GRISOGONO.

F.P. JOURNE

OCTA UTC

F.P. Journe innovates once again with the presentation of the Octa UTC, a patented system that mechanically indicates the different time periods linked with the earth's geographical positions, including summer and winter hours. On the dial, blue hands indicate the local hours, which are linked to the calendar indicating the current geographical time. The rose-gold hand indicates the hour of the second time zone on a 24-hour scale, represented by the earth's 24 time zones. The automatic-winding movement is constructed on the 1300.3 caliber, and offers an additional complication to the Octa line. The dial features off-centered hours, minutes and seconds, a retrograde power reserve, a large date, and for the first time, a dial with the earth divided in time zones. The passing of the months from 28 to 31 days is done manually.

FREDERIQUE CONSTANT

HEART BEAT MANUFACTURE GMT AUTOMATIC – REF. FC-938CDG4H6

The 42mm polished stainless steel case of this timepiece gently sets the tone for a silver dial of delicate intricacy. The Heart Beat aperture, iconic signature of the brand, at 6:00 breaks the tranquility of the Clous de Paris guilloché, revealing the inner workings of the FC-938 caliber with 42-hour power reserve, 28,800 vph, KIF antishock system, Glucydur balance, and bridges decorated with Côtes de Genève and perlage. Opposing the porthole, a 12:00 subdial expresses a second time zone with day/night indicator and Arabic numerals, countering the black applied Roman numerals of the outer chapter ring. In addition to the dial-side exhibition, this wristwatch boasts a transparent caseback secured with six screws.

GUY ELLIA

JUMBO HEURE UNIVERSELLE

One of the latest creations by designer Guy Ellia is the Jumbo Heure Universelle with a movement created by Frédéric Piguet. Caliber PGE 1150 features a 72-hour power reserve, a blue sapphire disc, and Côtes de Genève-finished bridges with rhodium plating. The day/night indicator, the grand date and 24 time-zone indications define its singular functions. The Jumbo Heure Universelle bears an exceptional 50mm size that is rarely seen in the world of luxury watches. Mounted on an alligator strap with a folding buckle, this model is presented in black gold and is also available in white or pink gold.

HARRY WINSTON

OCEAN DUAL TIME – REF. 400/MATZ44RL.W

Designed specifically for the world traveler, the Ocean Dual Time in 18K rose gold lets its wearer voyage through various time zones with ease. Powered by a self-winding movement with a 45-hour power resereve, the dial of this model is adorned in whites and blues with rose-gold hands that perfectly match the case. Its subdials represent home time and a second time zone, with night and day indicators, making it the perfect dual time zone watch.

HARRY WINSTON

OCEAN DUAL TIME BLACK EDITION – REF. 400/MATZ44ZKC.K2

The Ocean Dual Time Black Edition is as dark as the late hours of the night. Harder than titanium, and light as can be, the Zalium™ case is the perfect canvas upon which to execute this bold and new interpretation of the Ocean Dual Time. Coupled with a black opaline dial, this ultra-black watch boasts clear legibility and functionality, the most crucial components of any timepiece. Both the home time and second time zone subdials are adorned with SuperLumiNova, allowing both the local and home time to be easily seen by the wearer not only in total darkness, but against the dark color of the specially treated three-dimensional dial as well.

HARRY WINSTON

PREMIER EXCENTER TIME ZONE – REF. 210/MATZ41WL

Modern and refreshing, with an asymmetrical display and retrograde functions, the Premier Excenter Time Zone exemplifies the uncanny ability of Harry Winston to constantly innovate and set trends in an industry that has been around for centuries. The silver-toned, shot-peened and circular satin-brushed dial is a crisp and subtly elegant background for its off-centered functions. The principal time zone hours and minutes chapter ring surround the day/night indicator, while the large date is visible through a beveled aperture and the retrograde second time zone display is clearly highlighted. The 18K white-gold case features sharp angles and its iconic arches reflect those at the flagship Harry Winston salon in New York City. These arches merge harmoniously into the bezel, which radiates a contemporary architectural elegance matched by a seamless ergonomic feel.

HUBLOT

KING POWER UNICO GMT – REF. 771.CI.1170.RX

The first evolution to the exceptional Unico manufacture movement, the HUB1220 Unico base with GMT function is at the heart of this expressive timepiece with an innovative time zone complication. Controlled by a pusher at 2:00, four rotating aluminum discs simultaneously position themselves to display the time in one of 14 available cities. Within the 48mm micro-blasted black ceramic case, the black skeleton dial impressively promotes legibility, while the distinctive personality of the architecture, along with the watch's stunning mechanics, are made visible to the owner.

IWC

PILOT'S WATCH WORLDTIMER – REF. IW326201

This model is the ideal timepiece for the practical jetsetter. The Pilot's Watch Worldtimer aids its well-traveled wearer wherever he or she may be. Its 24-hour ring complete with clearly visible city names in white, as well as all 24-hour time zones, stands out against the blackened dial. It is quite simple for the owner of this piece to keep track of the various time zones across the world as the dial shows the current local time, yet can be advanced or set back by one hour as the owner voyages. This is even possible when crossing through the International Date Line. This timepiece is presented in a Ø 45mm stainless steel case complete with soft iron for protection against magnetic fields, as well as security against displacement by drop in air pressure. This model is truly perfect for the adventure seeker!

JACOB & CO.

GHOST

The Ghost timepiece from Jacob & Co. presents a new take on the GMT watch, with five mini-screens that present day, date, multiple world cities and battery reserve power. Ghost is powered by a lithium-polymer battery movement that beats at 32,768 Hz. A "mode" button allows the user to scroll through the watch's different screens, and a "color/GPS" button changes the color of the display with one short push (red, orange, yellow, green, blue, purple or white) or activates GPS localization, with more sustained pressure. The Ghost's battery can be charged via a USB cable hooked up to a computer or electrical outlet.

JACOB & CO.

GLOBAL 1

Bringing to life a rarely utilized pentagonal geometry, the Global features local time with the same dimensions as the other four time zones, making a powerful statement regarding the worldwide citizenship of its wearer. Individually powered by high-precision Swiss ETA quartz movements that provide hour and minute control for each time zone, the five dedicated crowns form a synergic emblem of global modernity. While 5 powerful variations each embrace an array of materials, designs and decorative techniques, the Global adorns the four additional dials of every model with the names of cultural and economic powerhouses Tokyo, New York, Paris and Los Angeles.

JAEGER-LECOULTRE

DUOMETRE UNIQUE TRAVEL TIME – REF. 606.35.2J

The Duomètre Unique Travel Time watch uses Jaeger-LeCoultre's groundbreaking Dual-Wing system for a unique world-time function. Two separate mechanisms power the piece, one for each complete time zone displayed. Governed by the same regulating organ, the mechanisms derive their energy from separate barrels, enhancing the watch's functioning and precision. In a first for world-time watches, the wearer can set the time in the second time zone to the nearest minute, thus allowing for exactitude in all spots upon the globe. A subdial at 2:00 presents the local time, while its companion at 10:00 displays the world-time hours and minutes. At 6:00, a miniature map also hosts the day/night indication and the hours disc. Within beats the Jaeger-LeCoultre Calibre 383, an exceptional movement that continues an illustrious tradition of immaculate hand-finishing.

PANERAI

LUMINOR 1950 3 DAYS GMT AUTOMATIC CERAMICA – REF. PAM00441

Combining the rich texture of its leather strap with the rugged charisma of its black ceramic case and bezel, this 44mm self-winding GMT timepiece leaves no doubt as to its masculine personality. Powered by the 229-component P.9001 caliber with two barrels providing three days of power reserve, the 28,800 vph watch displays, on its black dial with luminous enhancement, the date at 3:00, small seconds at 9:00 and a second time zone via a central hand with distinctive arrow-shaped pointer for ideal readability. Armed with a seconds-reset function for perfect time-setting synchronization, the timepiece displays its power reserve indicator on the transparent smoked sapphire crystal caseback.

PANERAI

RADIOMIR 3 DAYS GMT ORO ROSSO – REF. PAM00421

Marrying the geometric elegance of its classical Radiomir case with the luster of its warm, polished, red-gold allure, this hand-wound 47mm timepiece boasts a 210-component movement made entirely in house. Armed with two barrels, the P.3001 caliber permits a smooth choreography of small seconds at 9:00, subtle date at 3:00, and central GMT hand with arrow-shaped pointer. Displaying a second time zone with graceful refinement, this three-day power reserve timepiece is worn on a Panerai-personalized alligator strap with polished 18K red-gold buckle.

PANERAI

RADIOMIR 10 DAYS GMT ORO BIANCO – REF. PAM00496

Housed within a 45mm case in 18K white gold, this self-winding timepiece boasts a dial of rich informative personality and bright legibility. Demonstrating the date at 3:00, seconds and 24-hour indication at 9:00 and linear 10-day power reserve at 6:00, this three-barrel wristwatch presents its second time zone via a luminous central hand with arrow pointer for optimal clarity.

PANERAI

RADIOMIR 10 DAYS GMT ORO ROSSO – REF. PAM00497

Crafted in 18K red gold, this brown-dialed timepiece offers a warm depth of color as it showcases the sophistication of its 296-component P.2003 automatic caliber oscillating at a frequency of 28,800 vph. Within a highly intuitive architecture, the watch indicates the date through a 3:00 aperture, small seconds and 24h indicator via a 9:00 subdial, and second time zone centrally, distinguished by its arrow-shaped point. Completing the display at 6:00, the 10-day power reserve is read on a linear scale.

PANERAI

RADIOMIR 10 DAYS GMT PLATINO – REF. PAM00495

With its case and bezel in polished platinum, this 25-jewel GMT wristwatch reveals a window date, linear 10-day power reserve gauge, and small seconds/24h indicator. Joining the hour and minute hands on the black dial, a central arrow-shaped hand for the second time zone completes the choreography of the in-house P.2003 self-winding caliber with three barrels, Glucydur balance, and anti-shock device.

PATEK PHILIPPE

MEN'S COMPLICATIONS – REF. 5130/1G

Worn on a bracelet that reflects the gleam of its 39.5mm white-gold case, this world-time demonstration is powered by the 240 HU automatic movement with off-centered 22K gold mini-rotor. An explosion of sunburst guilloché on the gold dial sets the stage for its surrounding 24-hour scale with day/night indication. Encircling the expressive central design, 24 time zones, set via a pusher at 10:00, are illustrated on a sober and legible outer ring.

RICHARD MILLE

RM 022 TOURBILLON AERODYNE SECOND TIME ZONE

The RM 022 Tourbillon Aerodyne Second Time Zone is a visual and mechanical evolution of the first Richard Mille timepiece. Within its newly sculpted case design is the manual-winding caliber RM021, built on a baseplate of honeycombed orthorhombic titanium and carbon nanofiber. It provides hours, minutes, indicators for power reserve (circa 70 hours), torque indicator and function indicator for winding, neutral, and hand-setting operations. The finely finished pushbutton, located at 9:00, advances the second time zone in one-hour increments, utilizing a numbered, rotating disc of sapphire visible at 3:00.

RICHARD MILLE

RM 58-01 TOURBILLON WORLD TIMER JEAN TODT LIMITED EDITION

Designing a watch for world traveler Jean Todt, President of the FIA, Richard Mille wanted to create an easy and useful method for keeping track of travel through time zones with the RM 58-01. The time zone is set by simply rotating the bezel counterclockwise to set the name of the city where one has just landed to the 12 o'clock position, and the rest is adjusted automatically. The manual-winding tourbillon movement, with a baseplate and bridges in grade-5 titanium, has hours, minutes and a 10-day power reserve indication at 2 o'clock. A percentage of the profits from the sale of this exceptional watch will be donated by Jean Todt to the ICM (Institut du Cerveau et de la Moelle Épinière – Brain and Spine Institute) and to the FIA Action For Road Safety, whose goal is to support the United Nations' ten-year action plan for road safety (2011-2020).

VACHERON CONSTANTIN
OVERSEAS DUAL TIME – REF. 47450/B01R-9404

Dressed in a case and bracelet made of radiant 18K pink gold, the 1222 SC automatic movement beats in style at a rate of 28,800 vph. A pink-gold stripe at 3:00 ruptures the dial's fine finishing to indicate the remainder of the watch's 40-hour power reserve. Enhanced with luminous materials on the hours and minutes, the three-hand time display is joined by a date subdial at 2:00, and legible secondary time zone with black and white day/night indicator. On the watch's solid caseback, a stamped medallion of the Overseas sailboat tells the story of the collection's heritage.

VACHERON CONSTANTIN

PATRIMONY TRADITIONNELLE WORLD TIME – REF. 86060/000R-9640

Combining Vacheron Constantin's rich multiple time zones heritage with a driven spirit of innovation, the self-winding 2460 WT caliber takes into account the modern reality of the world's 37 time zones. By aligning the reference city with a black triangle at 6:00, the owner has immediate access to time in all of the globe's regions, including those with partial time zones, be they half or even quarter hours. Brilliantly accomplished within a 42.5mm 18K pink-gold case, this masterpiece ingeniously uses three dials to obtain such a supreme degree of complexity with optimal legibility. In addition to driving the central hours, minutes, and seconds, with a frequency of 28,800 vph, the movement precisely rotates a sapphire dial with 24-hour chapter ring and map-overlaying day/night indicator. Simplifying the use of the watch, all indication adjustments are made via a single crown. As if the front of the watch weren't spectacular enough, the Patrimony Traditionnelle World Time features a sapphire crystal openworked caseback.

ZENITH

CAPTAIN DUAL TIME – REF. 03.2130.682/02.C498

The Captain Dual Time model is equipped with a GMT hand with a different design, using a blue pointer that ensures optimal readability. This hand is set using a pushpin at 10:00. This mechanism makes it possible to adjust the second time zone with a simple gesture, without risking inadvertently changing the time on the watch. The Captain Dual Time is available in two versions, with a silver or black Clous de Paris guilloché dial, in a steel case, mounted on a metal bracelet or alligator leather strap.

Retrograde and Jumping Systems, Alternative Displays

Bell & Ross
Vintage WW1
Heure Sautante Platinum

The time is over
when watch dials were
nearly all equipped with
hands that rotated politely—

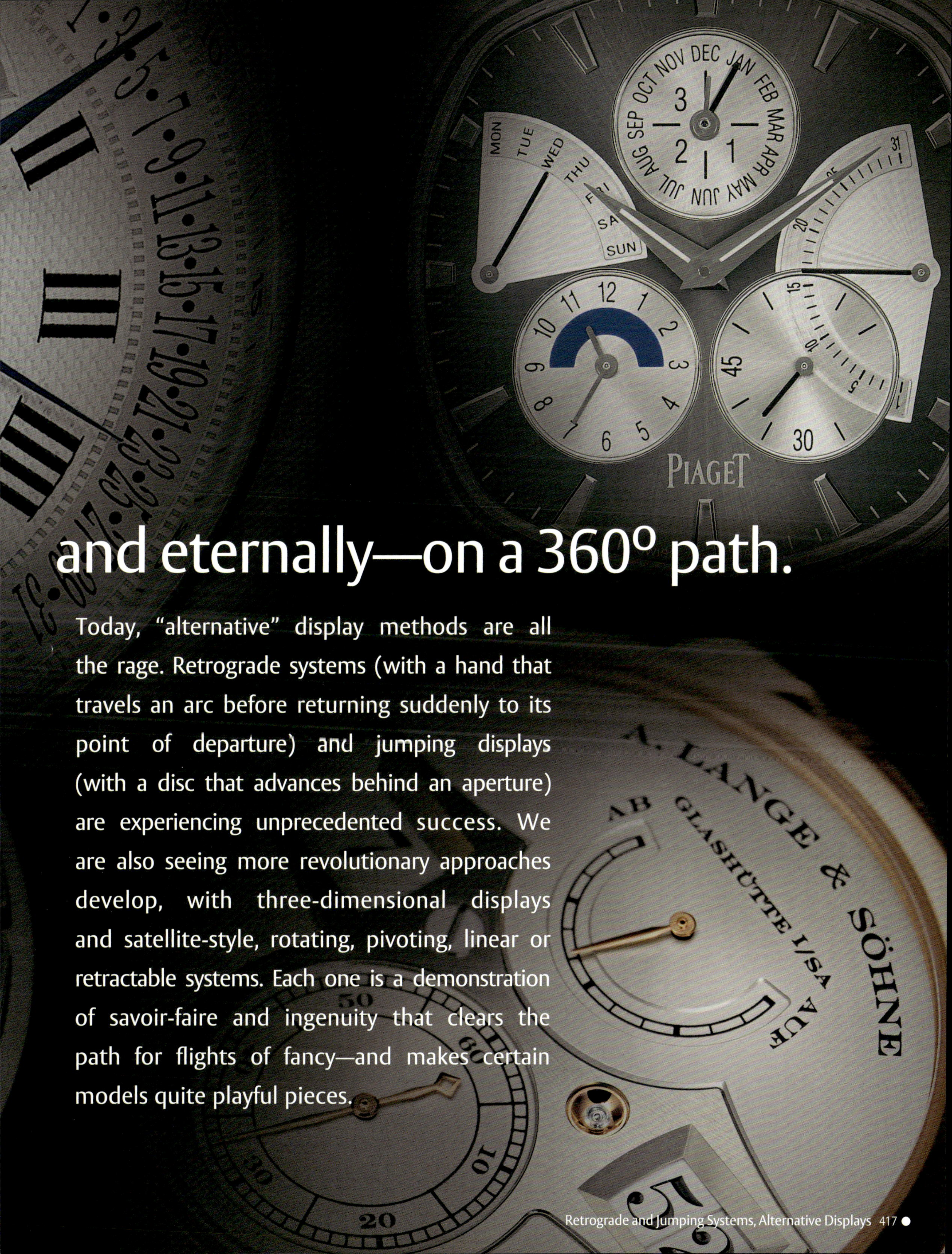

and eternally—on a 360° path.

Today, “alternative” display methods are all the rage. Retrograde systems (with a hand that travels an arc before returning suddenly to its point of departure) and jumping displays (with a disc that advances behind an aperture) are experiencing unprecedented success. We are also seeing more revolutionary approaches develop, with three-dimensional displays and satellite-style, rotating, pivoting, linear or retractable systems. Each one is a demonstration of savoir-faire and ingenuity that clears the path for flights of fancy—and makes certain models quite playful pieces.

Watchmakers have long sought to develop display methods that differ from the usual setup of hands turning... clockwise. This is why they invented the "retrograde" system (in which a hand moves across an arc, then instantaneously returns to its starting point to begin again) and the "jumping" displays (in which a disc moves behind an aperture). In the 18th century, Abraham-Louis Breguet enjoyed bringing life to his dials by using retrograde hands for certain indications, notably the date. From 1925-1935, watches with apertures on the dial enjoyed wide popularity. At the end of the 20th century, the mechanical renaissance brought with it a return to prominence of these alternative display methods, allowing watchmakers to vary the looks and styles of their creations.

TECHNIQUE

HOW DOES IT WORK?

RETROGRADE DISPLAYS

On a retrograde display, the hand traces an arc before returning in the blink of an eye at its original position to begin its path again. The construction of the base movement is the same as for an ordinary watch, and the difference lies in the display mechanism placed just below the dial. A complex system of springs, racks and pawls transforms the circular movement of the hand into a back-and-forth movement with ultra-fast return and immediate restart. The more retrograde displays are featured on a watch, the more precision and expertise are required to build it.

JUMPING DISPLAYS

On jumping mechanisms, the hand is replaced with a disc that "jumps" instantly with each change of unit. The base movement has the same construction as that of an ordinary watch, but a system of drawback and jumper springs serves to temporarily halt the indicator, accumulating energy until it is time to release it all at once to advance to the next increment. The jumping display is generally used for units lasting a significant amount of time, such as the hour or the date. It requires excellent control over energy, especially when several mechanisms of this type must spring into action at the same time.

Retrograde Displays

Retrograde displays are very fashionable at the moment, and their use is no longer limited to the minute and the date, instead spreading to indications such as the hour, seconds, day, chronograph counter, second time zone and other functions. The balletic movement of a hand springing back to its initial position often lends retrograde timepieces a highly spectacular appearances—especially when the dial combines two, three or even four of these indications, or when the watchmakers have a little fun by complicating the display even further. Another popular trend opens up the heart of the mechanism, allowing the wearer to peek "backstage."

Jean d'Eve
Sectora II Automatic
The Sectora II Automatic from Jean d'Eve displays the hour and minute across a 120° arc, housed in an asymmetrical case with a highly original shape.

Van Cleef & Arpels
Le Pont des Amoureux

Van Cleef & Arpels poetically reinvents the retrograde display on its model Le Pont des Amoureux, with a young woman's parasol marking the hours to the left, and her suitor's rose ticking off the minutes at right. The young lovers kiss twice a day during this continuous waltz.

Vacheron Constantin
Patrimony Contemporaine day-date bi-retrograde

The highly original Patrimony Contemporaine day-date bi-retrograde from Vacheron Constantin indicates the date on an arc from 9 to 3 o'clock, which is echoed by a second segment for the day from 7 to 5 o'clock.

Piaget
Emperador Coussin Perpetual Calendar

Equipped with an ultra-slim in-house movement, Piaget's Emperador Coussin Perpetual Calendar displays the day and the date using retrograde pointers on two fan-shaped segments respectively placed at 9 and 3 o'clock.

Longines
The Master Collection Retrograde Moon Phases

Equipped with an exclusive movement, The Master Collection Retrograde Moon Phases from Longines reveals no fewer than four retrograde displays—seconds, day date, 24-hour second time zone—on a barleycorn guilloché dial.

Milus, *Tirion Triretrograde*

The Tirion Triretrograde by Milus features a retrograde seconds display based on three 20-second segments. The first hand moves across its arc of a circle before moving instantly back to its starting point, while the second hand takes over, and then passes the baton on to the third.

Blancpain, *Villeret*

Upon its delicate guilloché dial, covered with translucent blue lacquer, this Villeret model from Blancpain displays a serpentine central date hand, as well as a retrograde seconds display on a 30-second segment at 6 o'clock.

Pierre Kunz, *Seconde Virevoltante Rétrograde*

On its Seconde Virevoltante Rétrograde, Pierre Kunz offers an amazing show of aerial acrobatics with a hand dipping and swooping around a 470° loop before returning to its initial position in 5/100ths of a second and starting all over again.

Christophe Claret, *X-TREM-1*

On the X-TREM-1 from Christophe Claret, the retrograde hour and minute hands are replaced by two small steel spheres placed in two sapphire tubes on either side of the case and which move using magnetic fields.

Pierre DeRoche
GrandCliff Double Retrograde Skyscrapers
Inspired by ever more daring, ever more vertiginous skyscrapers, Pierre DeRoche created the GrandCliff Double Retrograde Skyscrapers, with two vertical retrograde displays for the hours and the minutes.

Pierre DeRoche, *TNT Royal Retro Sapphire*
On the TNT Royal Retro Sapphire from Pierre DeRoche, equipped with an exclusive Dubois Dépraz mechanism, the seconds are indicated via a succession of six retrograde displays activated via a system of gears and springs that is visible on the dial.

Atlantic
Seaport Day Retrograde
The Seaport Day Retrograde from Atlantic displays the day with a hand that traverses an arc between 8 and 12 o'clock—completed by a double aperture for the large date at 1 o'clock and an oversized seconds subdial at 6 o'clock.

Jumping Displays

In jumping displays, the indication is provided by a disc on which are written numerals that "jump" by one increment at each change—as the date does on instantaneous calendars. This system is most often used for the hour display. However, horologers are increasingly intrigued by "digital" jumping indications, which they also use for other units (minutes or seconds), often revealing the discs upon the dials.

F.P. Journe, *Vagabondage II*

On the Vagabondage II from F. P. Journe, the only indication with a hand is the small seconds counter at 6 o'clock. Three discs appearing through a mostly openworked dial indicate the hours (at 12 o'clock) and the minutes (in the center) in two bisected apertures.

De Bethune, *Digitale*

The Digitale from De Bethune presents an entirely disc-based display, with an upper aperture for day, date and month, one in the center for the minutes and an aperture in the lower half of the dial for the hours.

Hysek, *IO Jumping Hour*

Outfitted with an exclusive movement and an atypical look, the IO Jumping Hour from Hysek presents a multi-level dial with jumping hours, dragging minutes displayed on a disc and day/night indicator.

RSW, *Outland*

On Outland from the brand RSW, the traditional hands make way for three concentric discs (hours, minutes and seconds) that move around a fixed point.

Vincent Calabrese, *Baladin*

Vincent Calabrese, on his Baladin, proposes an unusual way of reading the time by displaying the jumping hour in a small round window that marks off the minutes by rotating around the dial.

Harry Winston, *Opus 3*

For the Opus 3 by Harry Winston, Vianney Halter created an all-digital display with jumping hours, minutes, seconds and date appearing in six separate "portholes," and ten independently rotating discs driven by two mechanisms totaling 250 parts.

Cartier, *Rotonde de Cartier Jumping Hours*

A watch without hands, the Rotonde de Cartier Jumping Hours associates an aperture with jumping hours at 12 o'clock with a dragging minutes system equipped with a rotating disc. The hours are displayed in Roman or Arabic numerals.

A. Lange & Söhne, *Lange Zeitwerk*

The Lange Zeitwerk from A. Lange & Söhne associates traditional and digital displays, with a jumping hour on a double disc at 9 o'clock, jumping minutes at 3 o'clock and two counters with conventional hands for the small seconds and power reserve.

New Displays

Over the last few years, we have seen an explosion in the number of watches that revolutionize the indication of time, as well as its read-off. Whether three-dimensional, linear, pivoting, retractable or satellite systems, these alternative displays have become a favored field of exploration for young, creative watchmakers—who occasionally work behind the scenes with the biggest names in the industry. Though they do not always promise ease of reading, they are proof of a stunning ingenuity, served by unparalleled technical mastery.

Urwerk, *UR-1001 Zeit Device*

Urwerk's UR-1001 Zeit Device pocket watch is distinguished by its satellite system for the hours (with three rotating cubes) that also acts as a retrograde minute display, complemented by a satellite for months providing a simultaneous retrograde date display. The caseback also bears 100-year and 1,000-year indicators.

HYT, *H1 Black DLC*

The H1 from HYT—the "Hydro Mechanical Horologists"—scored a coup by introducing a retrograde hour display system using a fluorescent liquid propelled by the compression/decompression movements of two mini-reservoirs at 6 o'clock.

Ludovic Ballouard, *Half Time*

After the Upside Down, in which only the present time was displayed, Ludovic Ballouard continues along the path of original displays with Half Time. The hour appears at 12 o'clock with a Roman numeral emerging from the chaos, and the retrograde minute is displayed on an arc at 6 o'clock.

Pierre Kunz, *Infinity Looping*

On its Infinity Looping model, the Genevan watchmaker Pierre Kunz invented a spectacular new way of reading the time. Thanks to the gravitation of a satellite pinion that moves around a fixed rack, the hours and minutes whirl around the dial, tracing a rosette shape.

Cyrus, *Klepcys*

Created by the watchmaker Jean-François Mojon, the movement of the Cyrus Klepcys model displays the hour using a retrograde hand mounted upon two rotating cubes that indicate the day and night. The minutes and seconds are read off two central discs, and the date is indicated by a retrograde hand with a three-dimensional rotating tens column.

de Grisogono, *Meccanico dG*

The Meccanico dG dual time zone watch by de Grisogono combines an analog display of the first time zone with a digital display of the second, with both driven by a completely mechanical movement comprising 651 parts. The digital second time zone indication is composed of mobile micro-segments driven by 23 cams.

MCT, *Sequential One All Black*

The Sequential One by MCT puts on a fantastic show. The hour appears in turn in one of the four display zones on the dial by means of an original system of triangular prisms closely resembling animated advertising panels. The minutes appear over a 270° arc, and at each change of hour, the mobile minute ring makes a one-quarter turn to the left to reveal the following hour.

Cabestan, *Trapezium*

Equipped with a vertical tourbillon and a fusee-chain device, the highly original Cabestan Trapezium in titanium displays the hours, minutes, seconds and power reserve using a system of rotating drums.

Harry Winston, *Opus 7*

For Harry Winston, in the Opus 7, the watchmaker Andreas Strehler created an "alternating display." A first press on the crown guard turns the disc to bring the correct figure opposite the marker at 10 o'clock, a second press displays the minutes, and a third the power reserve.

Louis Vuitton, *Tambour Spin Time GMT*

The Tambour Spin Time GMT by Louis Vuitton marks the hours using 12 rotating rods arranged in a star shape, each tipped with a small cube. Every time the minutes move from 59 to 00, one of the cubes pivots to reveal the following hour, while the cube displaying the previous hour also makes a quarter-turn to reveal its neutral face.

Harry Winston, *Opus 8*

For the Opus 8 by Harry Winston, Frédéric Garinaud devised a system of digital hours and minutes display inspired by pin art games. Activating a bolt on the side of the case causes the digits corresponding to the time to "rise up" from the dial, while the minutes are marked off on a vertical scale.

Bulgari, *Chronographe Papillon*

On the Bulgari Chronograph Papillon from the Daniel Roth Collection, the minutes are displayed by an ingenious central device that completes a 360° rotation in two hours and boasts two independent pivoting hands that take turns traversing a 180° segment that marks 0 to 60 seconds.

Harry Winston, *Opus 9*

The Opus 9 by Harry Winston, developed by design engineer Eric Giroud and watchmaker Jean-Marc Wiederrecht, features a linear display of the hours and minutes by means of two small brass chains set with emerald-cut diamonds, along with vivid mandarin garnets serving as hour markers.

Urwerk, *UR-CC1 Black Cobra*

Inspired by the speedometers on certain 1950s cars, the UR-CC1 Black Cobra from Urwerk presents the linear display of the hours and minutes via two cylinders. Upon reaching the far end of its trajectory, the minute cylinder returns to its initial position in one tenth of a second, and this retrograde movement pushes the hour cylinder one unit forward.

Harry Winston, *Opus 12*

The Opus 12 from Harry Winston proposes an unusual way to read time, with 12 pairs of hands that pivot to mark the hours and five-minute increments. A central retrograde hand counts off the five minutes between the increments.

Chanel, *J12 Rétrograde Mystérieuse*

The central minutes hand of Chanel's J12 Rétrograde Mystérieuse runs in a traditional manner for the first ten minutes, before bumping up against the retractable crown appearing on the dial at 3 o'clock. This sets off a retrograde motion that takes another ten minutes to bring the hand to the other side of the crown, on the 20th minute.

Jean Dunand, *Palace*

For Jean Dunand's Palace model, Christophe Claret designed a flying tourbillon caliber with a second time zone and power reserve indication, situated in two vertical cartridges traversed by two ovals with arrows on them.

Christophe Claret, *DualTow*

On his DualTow watch (with tourbillon and single-pusher chronograph) Christophe Claret presents an hour and minute display by notched rubber belts inspired by caterpillar-track vehicles.

Harry Winston, *Opus 11*

Harry Winston's Opus 11, created by Dennis Giguet, has no hands at all. Every 60 minutes, the large central numeral for the hour explodes into chaos before being reconstituted in a fraction of a second, using 24 mobile "pallets."

Retrograde/Jumping Combinations

The jumping hour is often partnered with a retrograde minute, or other similar indications. This offers a fascinating spectacle on the hour and at midnight, when the numerals jump in their apertures and the hands leap to their starting position.

Hautlence, *Avant-garde HL[RS] 03*

The Avant-Garde HL[RS] 03 model from Hautlence associates a jumping hour and a retrograde minute, linked by an exclusive connecting-rod/crank system—all housed within an open-worked architecture that allows for beautiful 3D effects.

MB&F, *Horological Machine N° 2*

The Horological Machine N° 2 by MB&F stands out for its resolutely original two-dial look. The right-hand dial displays jumping hours and concentric retrograde minutes, while the left-hand provides retrograde date as well as twin-hemisphere moonphase displays.

de Grisogono, *FG One*

The FG One from de Grisogono boasts two dials, one with retrograde minute on a 230° arc, instantaneous jumping hour and dragging supplementary time zone, the other with a retrograde seconds complemented by a day/night aperture linked to the second time zone.

John Isaac, *Jumping Hour*

Developed in collaboration with Dubois Dépraz, the limited edition Jumping Hour from John Isaac stands out for its original design, which features an elliptical case framing a round subdial with jumping hours and retrograde minutes.

Bulgar

Octo Chronograph Quadri-Retro, Gérald Genta Collection

A true animated ballet, the Octo Chronograph Quadri-Retro, from Bulgari's Gérald Genta Collection, associates a jumping hour aperture with four retrograde indications: central minute, date at 6 o'clock and chronograph hour and minute counters at 3 and 9 o'clock, respectively.

A. LANGE & SÖHNE

LANGE ZEITWERK – REF. 140.025

The Lange Zeitwerk is powered by the manual-winding Lange proprietary L043.1 caliber, which contains 388 components and 68 jewels. Decorated and assembled by hand, the movement beats at 18,000 vph and possesses a precision beat-adjustment system with a lateral setscrew and whiplash spring. The dark tinted sapphire crystal glass has a special UV-transparent PVD coating that permits light to enter and charge the numerals with enough photonic energy to reliably emit the time for several hours during the night. The case is platinum and the series is limited to 100 pieces.

BELL & ROSS

WW1 HEURE SAUTANTE

The WW1 Heure Sautante is a Bell & Ross original complication for those who are passionate about timekeeping. The Arabic numeral hours are displayed through a large fixed window at 12:00, replacing hour hands. This aperture is vertically aligned with both the power reserve indicator at 6 and the central minute hand. The power reserve indicator is exceptionally important on this timepiece because the mechanism that triggers the jumping change in hours requires considerable energy. The wearer is thus made aware of how much power remains in the WW1 Heure Sauntante: essential information for seamless functioning of the movement.

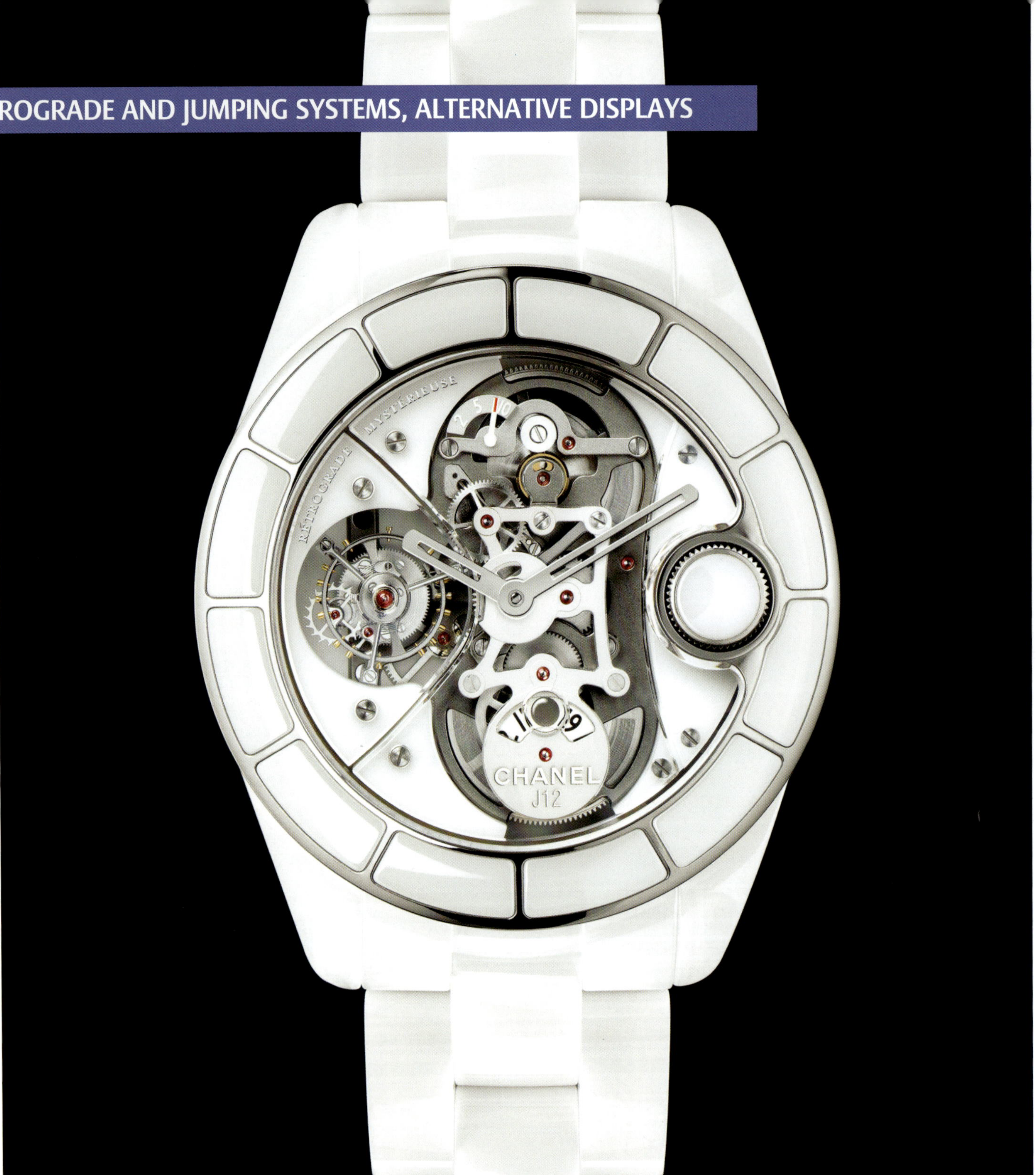

CHANEL

J12 RETROGRADE MYSTERIEUSE – WHITE CERAMIC

When technical prowess compliments the aesthetics
CHANEL RMT-10 Calibre

- R for Retrograde
- M for Mysterious
- T for Tourbillon

After introducing the first tourbillon featuring a highly innovative high-tech ceramic bottom-plate, CHANEL confirmed its status as a pioneer of contemporary watchmaking by creating the J12 Rétrograde Mystérieuse. The Rétrograde Mystérieuse is a concentrated dose of innovation, bringing together complications and a world première:

CHANEL

J12 RETROGRADE MYSTERIEUSE – BLACK CERAMIC

1. A tourbillon
2. A digital minutes display
3. A retrograde minutes hand
4. A 10-day power reserve
5. A retractable vertical crown

This perfectly round watch with a 47mm diameter has been designed without a side crown to ensure optimum comfort on the wrist. The work on this complication was entrusted to one of the most state-of-the-art watchmaking design/construction workshops: the Giulio Papi team (APRP SA). The piece is available in a limited and numbered edition of ten pieces in black high-tech ceramic and 18K white gold or ten pieces in black high-tech ceramic and 18K pink gold. The white high-tech ceramic edition is a one-of-a-kind piece.

CHRISTOPHE CLARET

X-TREM-1 – REF. MTR.FLY11.040-048

X-TREM-1, the first of a new generation of timepieces, represents a major technical and aesthetic accomplishment: a flying tourbillon inclined at a 30° angle, mounted on a three-dimensional curvex titanium mainplate, equipped with a retrograde hours and minutes display system that veers radically from existing watchmaking conventions. Two tiny hollowed steel spheres, isolated within sapphire tubes on the left and right sides of the caseband, magically move with no mechanical connection, thanks to magnetic fields. The hand-wound watch draws its energy from two barrels—a necessity in this case, since they enable the use of a sophisticated display without disturbing the rate of the tourbillon, and thus the accuracy of the watch.

HARRY WINSTON

HARRY WINSTON OCEAN TRIPLE RETROGRADE CHRONOGRAPH REF. 400/MCRA44RC.K

The Ocean Triple Retrograde Chronograph is an extremely intricate and alluring timepiece. Where most modern chronographs present time on circular registers, this unique model presents displays time on three separate retrograde counters, each of which—as it reaches the limit of its arc—must fly back immediately to its starting position so the interval timing can continue onward. This timepiece achieves the remarkable feat of extreme accuracy through thorough and painstaking craftsmanship. Extra energy is required for each of the model's three hands to fly back to its starting point. A supremely sophisticated piece and a true mechanical marvel, Harry Winston's Ocean Triple Retrograde Chronograph shines inside a brilliant rose-gold case.

HARRY WINSTON

HARRY WINSTON OCEAN TRIPLE RETROGRADE CHRONOGRAPH REF. 400/MCRA44ZKL.S

Harry Winston proves to be visionary time and time again. The Ocean Triple Retrograde Chronograph in Zalium™, a material unique to Harry Winston in watchmaking, exemplifies the house's mastery not only concerning the inner workings of the watch, but also in the fact that its harder-than-titanium, non-allergenic and anti-corrosion case is composed of a material developed by the geniuses at the house themselves. The Zalium™ DLC-coated case frames a silver vertical-brushed finish with silver applied subdial and counters. The functions of this watch are extremely unique as they display time on three separate retrograde counters rather than on traditional circular chronograph subdials. Harry Winston proves to be audacious as well as meticulous with this truly special model.

HAUTLENCE

AVANT-GARDE HL^{RQ} – REF. HL^{RQ}02

The Avant-Garde HL^{RQ} bears the trademarks of Hautlence's style: a house movement that powers a specific time display with jumping hours, retrograde minutes and instantaneous calendar. It incorporates the codes of previous Hautlence collections, reinterpreting them for a modern style befitting its place in the house's Avant-Garde family. The piece's 44mm case, crafted in steel and titanium, boasts larger, more imposing horns than its predecessors. The sapphire dial—designed by the brand—lends transparency and depth, revealing the in-house mechanical hand-wound movement with its jumping hours and retrograde minutes (using the Hautlence connecting rods system) and its calendar. The impression of weightlessness is further enhanced by the overlapping discs, which created a stepped effect. This ultra-modern lifestyle piece is worn on an alligator leather strap with titanium folding or pin buckle.

HAUTLENCE

AVANT-GARDE HL^{RS} – REF. HL^{RS}03

Created in Hautlence's Atelier d'horlogerie contemporaine, the Avant-Garde HL^{RS} offers an innovative approach to time display. The two multilayered dials create an exquisite play of depth and transparency that reveals the architectural balance of the piece. The unique transparent sapphire dial's upper face features minutes and indexes; its lower face includes a semi-opaque area and a coating made from highly polished rhodium. A honeycombed base dial further enhances the feeling of weightlessness. Luminescent numerals and indexes reinforce the three-dimensional effect of the jumping hours, retrograde minutes and small seconds. An ergonomic case in steel and titanium, covered by black DLC, imparts strength and lightness to the piece, which incorporates all of the brand's characteristic elements: in-house manual-winding movement, jumping hours and retrograde minutes using Hautlence's connecting rods system, hand-drawn strokes and hand-angled finish.

VACHERON CONSTANTIN

PATRIMONY CONTEMPORAINE RETROGRADE DAY AND DATE – REF. 86020/000G-9508

Powered by the mechanical automatic-winding Vacheron Constantin Caliber 2460 R31 R7, this Patrimony Bi-Retrograde is hallmarked with the Poinçon de Genève. Beating at 28,800 vph and offering 43 hours of power reserve, the 27-jeweled movement is equipped with a guilloché rotor. Retrograde day is provided at 6:00 and retrograde date is shown at 12:00. The three-piece case is polished and water resistant to 3atm. The sapphire caseback reveals the exquisite movement.

Calendars

IWC
Big Pilot's Watch
Perpetual Calendar Top Gun

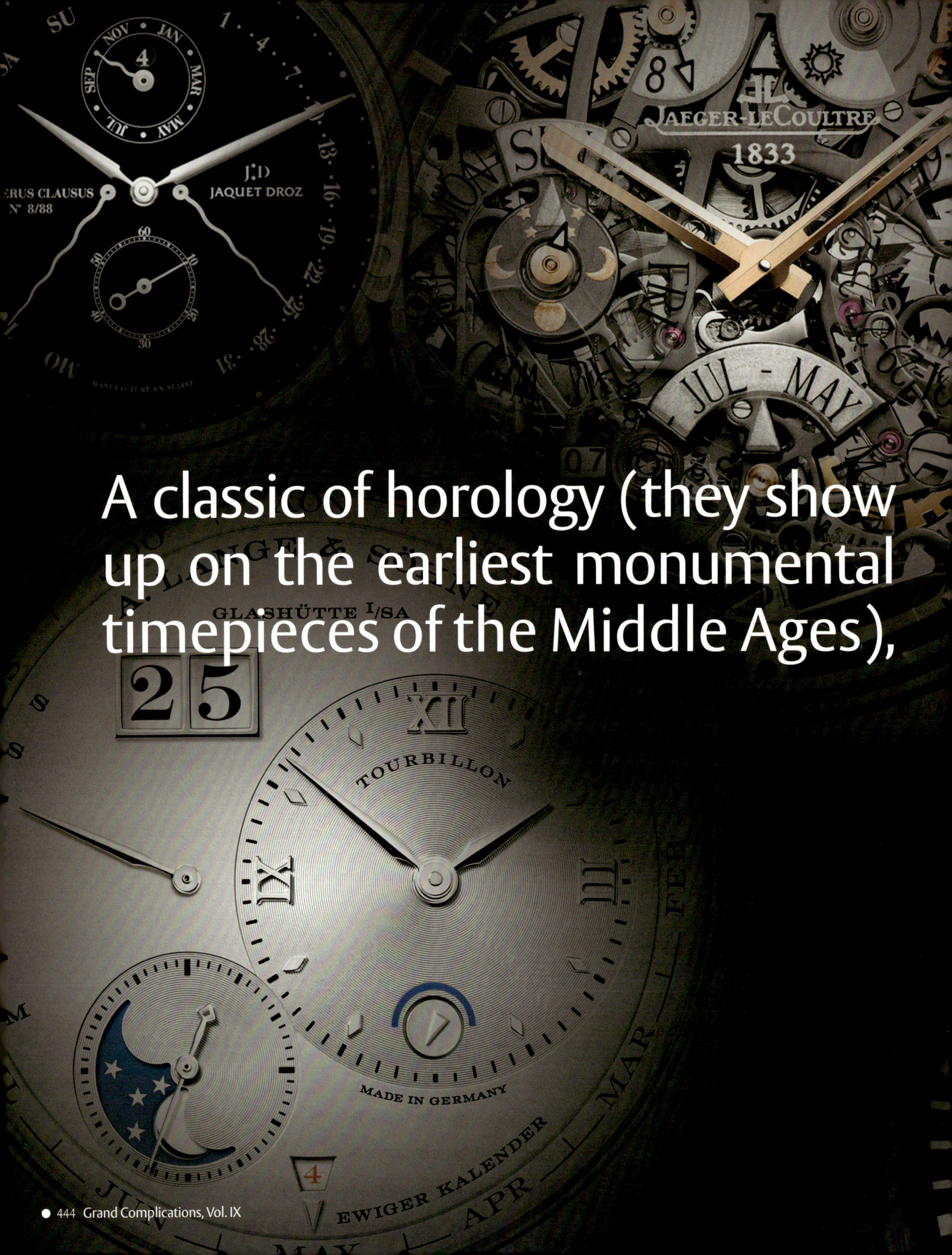
A classic of horology (they show up on the earliest monumental timepieces of the Middle Ages),

Movado, *Red Label Museum Calendomatic*

Inspired by a model launched in 1946, the Red Label Museum Calendomatic by Movado combines the famous understated black Museum dial with an original date display by means of a rotating disc bearing red Roman or Arabic numerals.

Hermès, *Le Temps Suspendu*

Le Temps Suspendu, from Hermès, is pure magic. With a simple press on the pushbutton, the hour and minute hands stand still next to 12 o'clock, while the date display disappears. Another press and all three hands return to their proper places.

Longines, *Master Collection Retrograde*

A festival of retrograde displays enlivens the Master Collection Retrograde by Longines, showing the date between 1 and 5 o'clock and the date at 12 o'clock, complemented by equally retrograde dual time zone and small seconds indications.

Jaquet Droz, *Grande Seconde Quantième*

On the Grande Seconde Quantième from Jaquet Droz, the date display is effected with a hand on the oversized seconds subdial at 6 o'clock, while hours and minutes share the off-centered subdial at 12 o'clock.

Cuervo y Sobrinos, *Robusto Day-Date Churchill*

The Swiss-Cuban brand Cuervo y Sobrinos has dedicated this limited edition, with aperture date and day displays, to one of the most famous cigar-smokers of all time.

Dubey & Schaldenbrand
Lady Grand Dome

On the tonneau-shaped Lady Grand Dome from Dubey & Schaldenbrand, the chronograph indications along the vertical axis meet a calendar with a central date hand and a double day/month aperture at 12 o'clock.

Blancpain, *Retrograde Calendar*

The Retrograde Calendar from Blancpain's women's collection displays the date upon its mother-of-pearl dial using a sinuous hand tipped with a small star. It is also equipped with a moonphase and a flower-shaped oscillating weight.

Perpetual Calendars: Classic Designs

Representing the ultimate form of calendar watch, the famous perpetual calendars are characterized by their multiple displays (date, day, month, leap-year cycle), generally enriched by moonphases. In their most classic version, they generally feature three or even four sub-dials arranged symmetrically in relation to the dial enter. But this restrained elegance sometimes conceals technical developments that considerably enhance the user-friendliness and the security of these ultra-sophisticated mechanisms.

Audemars Piguet
Jules Audemars Perpetual Calendar
Equipped with an ultra-thin self-winding caliber, the Jules Audemars Perpetual Calendar model by Audemars Piguet is a perfect example of a classic and yet timeless design. The leap-year cycle appears in the center of the month subdial.

Blancpain
Villeret Traditional Chinese Calendar
To appeal to Chinese watch collectors, Blancpain has released the first wristwatch equipped with a traditional Chinese (lunisolar) calendar, with a "double-hour" indication, Zodiac signs, Chinese month and date, the five elements, the ten "celestial trunks" and leap months.

Patek Philippe
Ladies First Perpetual Calendar, Ref. 7140
With its Ladies First Perpetual Calendar in rose gold, Patek Philippe presents its ultra-thin perpetual calendar—a classic from the manufacture—in a women's version for the first time.

Jaeger-LeCoultre
Master Grande Tradition Tourbillon à Quantième Perpétuel
Jaeger-LeCoultre has combined two major complications in its Master Grande Tradition Tourbillon à Quantième Perpétuel, featuring three subdials with date/day/month subdials, as well as the four-digit year number appearing in the center of the month window.

Blancpain, *Perpetual Calendar with Under-Lug Correctors*
On its Perpetual Calendar with Under-Lug Correctors, Blancpain has developed an original system serving to integrate the correctors within the four bracelet lugs, on the wrist side. A press of the finger is enough to correct the date, day, month, leap-year cycle or moonphase display.

IWC
Big Pilot's Watch
Perpetual Calendar Top Gun
Dedicated to the famous elite fighter pilot school, the Big Pilot's Watch Perpetual Calendar Top Gun houses its sophisticated movement within a high-tech ceramic case. The four-digit year display appears in an aperture at 7:30.

Perpetual Calendars: New Configurations

The design of perpetual calendars enables brands to play with various aesthetic arrangements in order to make their models stand out from traditional dials. The trend is towards a number of apertures placed either in a straight line or around the arc of a circle. Watchmakers also enjoy combining pointer- and window-type displays in sometimes highly original compositions, a sophisticated aesthetic approach that is associated with a number of technical accomplishments.

Ateliers deMonaco
Quantième Perpétuel QP-EZ

On its Quantième Perpétuel QP-EZ (EZ for "easy"), Ateliers deMonaco—the high-end division of Frédérique Constant—offers an ultra-simple setting system that allows the wearer to correct each of the three calendar displays independently using the crown.

A. Lange & Söhne
Lange 1 Tourbillon Perpetual Calendar

The Lange 1 Tourbillon Perpetual Calendar from A. Lange & Söhne presents its different indications in a very original way: a small, off-center hours/minutes subdial, large date in a double aperture, day with a retrograde hand at 9 o'clock and month on a disc that turns around the dial.

Antoine Martin
Quantième Perpétuel QP01

Equipped with a silicon escapement that needs no lubrication, the Quantième Perpétuel QP01 from Antoine Martin stands out for its original aesthetic, with vertical displays for the day and month, large date and "new-style" guillochage.

F. P. Journe
Octa Perpétuelle

The extremely original dial of the Octa Perpétuelle by F.P. Journe, with its offset hours, minutes and seconds, gives a starring role to the central retrograde hand, along with two day and month windows.

H. Moser & Cie., *Perpetual 1*

H. Moser & Cie. offers an unusual and extremely restrained take on the perpetual calendar. The month is displayed by a small central hand pointing to one of the 12 hour markers—such as 5 standing for May. The calendar may be adjusted forwards or backwards at any time.

Breguet, *Classique 7717BA*

On this self-winding Classique watch by Breguet, the pointer- or window-type displays are lined up along the same vertical axis, against the brand's characteristic guilloché dial.

Perpetual Calendars: Retrograding and Roller Systems

How can one enliven a perpetual calendar model? By choosing a retrograde-type mechanism for one or several of the calendar indications, with a hand moving across the arc of a circle before jumping back to its point of departure (see the chapter on Retrograde Displays and Jumping Hours, Alternative Displays). Another, much rarer, option is that of roller-type indications. This quest for originality paves the way for all manner of fanciful visual touches calling for a superlative degree of technical mastery.

Jaquet Droz, *Quantième Perpétuel*

Jaquet Droz, known for its consistently original designs, offers a Perpetual Calendar model with double retrograde display occupying the entire upper part of the dial, with days on the left and the date on the right.

Jean Dunand, *Shabaka*

The ultra-sophisticated Shabaka watch by Jean Dunand is equipped with a minute repeater striking on a cathedral chime, and is also distinguished by its roller-type display of the perpetual calendar indications, all of which jump instantly at midnight.

Piaget, *FortyFive Perpetual Calendar*

On its FortyFive Perpetual Calendar, Piaget presents a dynamic configuration, with retrograde date at 3 o'clock, retrograde day at 9 o'clock and subdial displaying the month and leap-year cycle at 12 o'clock.

Parmigiani
Toric Corrector

Equipped with a minute repeater and a pusher enabling instant correction of all the calendar and moonphase functions, the Toric Corrector by Parmigiani displays the retrograde date on the arc of a circle running from 8 to 4 o'clock.

Maîtres du Temps, *Chapter Two*

The Chapter Two watch, created by Daniel Roth, Roger Dubuis and Peter Speake-Marin for the Maîtres du Temps brand, displays the months and days spelled out on two rollers—complete with two correctors housed on the caseback.

Perpetual Calendars: The Transparent Approach

Visible mechanisms are all the rage in the watch industry at the moment. After introducing sapphire crystal casebacks on virtually all their complicated models, brands are vying with other to find ingenious means of revealing all or part of the movement through the dial side. Perpetual calendars are no exception, and there is a vast range of models with openworked dials, skeletonized movements or transparent dials that now hide almost nothing of their inner mysteries, at least in visual terms. These spectacular stagings are often matched by some exceptional technical feats accomplished behind the scenes.

Patek Philippe
Grand Complication Ref. 5104
On the Grand Complication Ref. 5104 model by Patek Philippe, the days, months and the leap-year cycle number are printed directly on the sapphire crystal dial and dark "pallets" highlight the appropriate white transfer by contrast.

Jaeger-LeCoultre
Master Eight Days Perpetual SQ
The Master Eight Days Perpetual from Jaeger-LeCoultre, endowed with an eight-day power reserve, is shown here in a dazzling skeletonized version featuring daringly curved bridges echoing the lines of longitude and latitude.

Vacheron Constantin
Malte Retrograde Perpetual Calendar Openface
Vacheron Constantin has equipped its Malte Retrograde Perpetual Calendar Openface watch in platinum with a double sapphire crystal. Transparent subdials with white rotating "pallets" serve to indicate the day, month and leap-year cycle painted in black on one of the sapphire crystals.

Cartier
Tortue Perpetual Calendar
Equipped with an in-house movement, the Tortue Perpetual Calendar by Cartier combines various types of display—including a central pointer-type date indication and a retrograde day running from 8 to 4 o'clock—appearing on an openworked dial.

Parmigiani
Tonda 42 Retrograde Perpetual Calendar
On the Tonda 42 Retrograde Perpetual Calendar by Parmigiani, the abbreviations of the day and month appear on two discs featuring original shapes and visible through the transparent dial.

Large Dates

Highly prized for their readability as well as their original touch, "large date" displays are becoming increasingly sought-after in recent years, whether appearing alone on the dial or associated with other complications. Several well-known brands have taken an interest in this ingenious mechanism, improving both its running and its reliability.

A. Lange & Söhne, *Datograph Up/Down*

Large date displays are closely associated with A. Lange & Söhne. A specialist in this type of display, the German brand uses its patented system on several models, including the Datograph Up/Down chronograph, with flyback function and power reserve indicator.

A. Favre & Fils

Phoenix 10.1 Quantième à Grand Affichage Rotatif

On the Phoenix 10.1 Quantième à Grand Affichage Rotatif by A. Favre & Fils, the units numerals remain immobile, while a red arrow (for the dates from 1 to 9) or the 1, 2 or 3 o'clock numeral move to indicate the date.

H. Moser & Cie, *Monard Date*

Based on two superimposed discs, the large date display on the Monard Date by H. Moser & Cie is also distinguished by its patented Double Pull Crown system, which avoids the user accidentally changing the time when simply seeking to adjust the date.

Cecil Purnell, *Grande Date*

The Grande Date from Cecil Purnell uses transparent effects, revealing the two discs of the oversized date display, as well as part of the movement and the tourbillon carriage.

Carl F. Bucherer, *Patravi EvoTec BigDate*

Designed for feminine wrists, the Patravi EvoTec BigDate from Carl F. Bucherer displays its large date in a double aperture situated in the upper left corner of the dial. The small seconds subdial at 6 o'clock echoes the cushion shape of the case.

Audemars Piguet

Tourbillon Grande Date Edward Piguet

Combining two superimposed discs and an ingenious system of partial gears, the display created by Audemars Piguet is intended to be the largest on the market. Witness this ultra-sophisticated Edward Piguet Tourbillon Large Date model.

Annual Calendars

The annual calendar continues its spectacular breakthrough. Less complex than a perpetual calendar, it enables brands to create watches that are more affordable and simpler to operate, while providing a degree of user-friendliness far superior to simple calendars in that they require just one correction per year. The last few years have brought a flourishing spate of annual calendar models featuring a great variety of displays and designs, as well as occasionally associating this complication with other useful functions.

Rolex

Oyster Perpetual Sky-Dweller

Rolex's travel watch Oyster Perpetual Sky-Dweller pairs a dual time zone with an annual calendar. The month is displayed discreetly in 12 apertures arranged around the dial. Setting the time zones and calendar is done using the crown.

Patek Philippe

Annual Calendar Regulator Watch, Ref. 5235

Launched in 1996, the patented Annual Calendar from Patek Philippe has become a bestseller for the brand. It has been interpreted in several different models, including the Annual Calendar Regulator Watch, with escapement and balance spring in Silinvar, a silicon derivative.

Cartier

Rotonde de Cartier Annual Calendar

The Rotonde de Cartier Annual Calendar is equipped with a large date in a double aperture at 12 o'clock and a rapid-setting system for the calendar, governed by the crown. The days and months are indicated on two concentric rings by two hands of which only the red tips are visible.

Girard-Perregaux

Cat's Eye Annual and Zodiac Calendars

Designed for women, the Cat's Eye Annual and Zodiac Calendars watch combines a snail-shaped date display with a large comet's tail-shaped aperture revealing the ever-circling dance of the Zodiac signs.

Parmigiani, *Tonda Retrograde Annual Calendar*

On Parmigiani's Tonda Retrograde Annual Calendar, the date is displayed using a retrograde hand traversing an arc from 8 to 4 o'clock, while the day and month (as a number from 1-12) appear in two small apertures.

A. Lange & Söhne
Saxonia Calendrier Annuel

The first A. Lange & Söhne watch to be equipped with an annual calendar, the Saxonia Annual Calendar combines the German brand's celebrated and patented large date display with three classic windows for the day, month and moonphase.

Omega
De Ville Hour Vision Annual Calendar

Equipped with a self-winding movement, a co-axial escapement and an annual calendar with instant date jump, the Omega De Ville Hour Vision Annual Calendar displays the month and date in a double window at 3 o'clock.

Ochs und Junior, *Anno*

The epitome of minimalism, the contemporary Anno from Ochs und Junior displays the calendar indications using three circular scales: date (1 to 31) around the edges of the dial, day (1 to 7) on the central axis, month (1 to 12) above the hour marker at 6 o'clock. The annual calendar mechanism designed by the famous Ludwig Oechslin has only three moving parts—as opposed to 40 in a conventional system.

Hermès, *Dressage Annual Calendar*

The Dressage Annual Calendar watch by Hermès, with its exclusive Vaucher Manufacture Fleurier movement, is distinguished by its pointer-type 270° retrograde date display and its two visible day and month discs.

Hysek, *Kilada Annual Calendar*

Combining rose gold and black rubber in a highly contemporary design, the Kilada Annual Calendar possesses an annual calendar with a month counter at 8 o'clock and a large date at 12 o'clock.

Pierre DeRoche
GrandCliff QA Power Reserve

The GrandCliff QA Power Reserve model by Pierre DeRoche, equipped with a Dubois Dépraz movement, combines the annual calendar with a chronograph featuring retrograding 60-minute and 6-hour counters, a flyback function and a large date.

Zenith, *Captain Winsor Annual Calendar*

Equipped with the automatic-winding El Primero caliber, Zenith's Captain Winsor Annual Calendar associates the functions of a 1/10-of-a-second chronograph with an annual calendar boasting three apertures for the date, day and month.

A. LANGE & SÖHNE

DATOGRAPH PERPETUAL – REF. 410.032

This model is driven by the manually wound Lange L951.1 movement. It is a flyback chronograph with a precise jumping minute counter as well as perpetual calendar with a large date, moonphase display, day of the week, month, leap-year display, small seconds hand with stop seconds, and day/night indicator. The plates and bridges are made of silver and the balance cock is engraved by hand. The three-piece rose-gold case features an untreated German antireflective sapphire crystal and caseback, as well as a main pushbutton for simultaneously advancing all calendar displays, recessed pushbuttons for separately advancing the calendar displays and two chronograph pushbuttons. The hand-stitched crocodile strap is secured with a solid gold buckle.

A. LANGE & SÖHNE

GRAND LANGE 1 – REF. 117.028

With a sleek black solid silver dial and the classic Grand Lange 1 dial configuration framed by a thinner white-gold case, this piece is driven by the manufacture caliber L095.2, whose 72-hour power reserve is provided by a single mainspring barrel, reducing the thickness of the iconic piece. Rhodiumed gold appliqués on the dial, paired with luminescent hands, make this Grand Lange 1 highly legible at all times of day or night. The movement boasts several manufacture touches, such as a classic screw balance, a balance spring developed and produced in-house, a three-quarter plate in untreated German silver and lavishly hand-decorated components.

A. LANGE & SÖHNE

GRAND LANGE 1 "LUMEN" – REF. 117.035

The Lange outsize date, combined with an off-centered dial, became an emblem of the manufacture soon after its introduction. The technical secret to such an imposing indication, however, remained hidden behind the solid silver dial...until now. Parts of the dial have been replaced by semi-transparent sapphire crystal—only the outer ring and the subdials for seconds and hours and minutes are in blackened silver—revealing the disc mechanism behind the date indication. The urge to show all pervades the Grand Lange 1 "Lumen": in addition to the transparent dial, the piece also boasts luminescent indications for easy night readoff, as well as a sapphire crystal caseback that exposes the manual-winding manufacture caliber L095.2. The timepiece is available in a limited edition of 200 watches in platinum cases.

A. LANGE & SÖHNE

LANGEMATIK PERPETUAL – REF. 310.025

Powered by the self-winding Caliber L922.1 SAX-0-MAT, the Langematik Perpetual is the world's first self-winding wristwatch with a large date and a perpetual calendar featuring separate pushbuttons for individual corrections of the calendar indications (date, day of the week, month, leap year, and moonphase displays), as well as a main pushbutton for advancing all displays. Caliber L922.1 SAX-O-MAT has a 46-hour power reserve, stop seconds with a patented zero-reset mechanism, plates and bridges made of untreated German silver, and a hand-engraved balance cock. The Langematik Perpetual's three-piece platinum case is fitted with antireflective sapphire crystal on the front and back, and a crown for winding the watch and setting the time. The solid silver and argenté dial is topped with luminous platinum hands. The watch's crocodile strap is secured via a solid gold buckle.

BLANCPAIN

VILLERET 8 DAY COMPLETE CALENDAR – REF. 6639-3437-55B

Presented in platinum with a black enamel dial, the Villeret 8 Day Complete Calendar displays a moonphase with small seconds at 6:00, date via a central hand against an outermost date ring, and day and month through nicely contrasting apertures at 12:00. Powered by the 6639 self-winding caliber with eight-day power reserve, this watch features an exhibition caseback and is worn on a black alligator leather strap that complements the deep tone of the dial.

BLANCPAIN

VILLERET CHINESE CALENDAR – REF. 00888-3413-55B

On the fascinating dial of Blancpain's Villeret Chinese Calendar, the hours, minutes and Gregorian calendar rub shoulders with the main indications of the Chinese calendar: traditional double-hour indication, day, month with indication of leap months, and signs of the Chinese Zodiac, as well as the five elements and the 10 celestial stems. The self-winding Calibre 3638 powering this watch is endowed with a 7-day power reserve, despite the presence of 434 parts including 39 jewels. Its complexity stems as much from the number of indications to be displayed as from the irregular nature of their cycles. The 45mm platinum case is fitted with a crown adorned with a cabochon-cut ruby and comprises five integrated under-lug correctors serving to adjust the indications.

DE BETHUNE

DB25QP – REF. DB25QPR

A poetic tale of horological finesse and technical savoir-faire, this 44mm timepiece's silver-toned, hand-guilloché dial expresses the brilliance of the perpetual calendar with radiance and captivation. Joining apertures at 3:00 and 9:00, where the month and day may be read with perfect clarity, a 6:00 date subdial echoes the circular splendor of the watch's visual centerpiece at 12:00. Playing off the hand-polished, flame-blued steel sheen of the three revolving hands, the 47-jewel DB 2324 QP automatic caliber's spherical moonphase indicator in platinum and blued steel revolves on its axis within a sky adorned with gold stars and a discreet leap-year indicator window.

F.P. JOURNE

QUANTIEME PERPETUEL

The ninth creation of the Octa collection, the new Quantième Perpétuel presents a distinctive aesthetic through an uncluttered dial with apertures for the day, date and month. The instantaneous jump is reinforced by an ingenious F.P. Journe system designed to accumulate energy, release it instantaneously when the date, day or month changes, then slow it down at the end of its route. The varying lengths of the months are automatically taken into account, and the leap-year cycle is displayed at the center of the dial. The automatic-winding movement, constructed on the basis of the exclusive Octa calibre 1300.3, is manufactured in 18K rose gold and possesses 120 hours of power reserve. The Quantième Perpétuel is available in platinum or in 18K red gold, with 40mm or 42mm diameter.

FRANCK MULLER

CINTRÉE CURVEX PERPETUAL CALENDAR – REF. 9880CCQPB5NE

A poetic tale of horological finesse and technical savoir-faire, this 44mm timepiece's silver-toned, hand-guilloché dial expresses the brilliance of the perpetual calendar with radiance and captivation. Joining apertures at 3:00 and 9:00, where the month and day may be read with perfect clarity, a 6:00 date subdial echoes the circular splendor of the watch's visual centerpiece at 12:00. Playing off the hand-polished, flame-blued steel sheen of the three revolving hands, the 47-jewel DB 2324 QP automatic caliber's spherical moonphase indicator in platinum and blued steel revolves on its axis within a sky adorned with gold stars and a discreet leap-year indicator window.

FRANCK MULLER

LONG ISLAND PERPETUAL CALENDAR – REF. 1350 T QP

The ninth creation of the Octa collection, the new Quantième Perpétuel presents a distinctive aesthetic through an uncluttered dial with apertures for the day, date and month. The instantaneous jump is reinforced by an ingenious F.P. Journe system designed to accumulate energy, release it instantaneously when the date, day or month changes, then slow it down at the end of its route. The varying lengths of the months are automatically taken into account, and the leap-year cycle is displayed at the center of the dial. The automatic-winding movement, constructed on the basis of the exclusive Octa calibre 1300.3, is manufactured in 18K rose gold and possesses 120 hours of power reserve. The Quantième Perpétuel is available in platinum or in 18K red gold, with 40mm or 42mm diameter.

GLASHÜTTE ORIGINAL

SENATOR PERPETUAL CALENDAR – REF. 100-02-22-05-05

Made by hand in the Glashütte manufacture, the Senator Perpetual Calendar exemplifies the ongoing quest for the perfect union of mechanical complications and art. All Senator Perpetual Calendar displays are arranged to form a clear, harmonious balance. The day, month and Glashütte Original Panorama Date display black numerals against a white background, and the warm silver dial shows a silver moonphase display on a blue disc. A discreet, unambiguous leap-year display presents a red dot to indicate a leap year, followed by yellow, black and white dots for successive years. The heart of the Senator Perpetual Calendar, visible through the sapphire crystal caseback, is the Glashütte Original Caliber 100-02, a 28,800 vph automatic movement featuring a screw balance with 18 weighted golden screws and a power reserve of more than 55 hours.

GLASHÜTTE ORIGINAL

SENATOR PERPETUAL CALENDAR – REF. 100-02-25-12-05

On this stainless steel, black-dial version of the Senator Perpetual Calendar, the moonphase appears on a black disc adorned with silver stars and a radiant silver moon. The timepiece's reset mechanism allows easy synchronization of the second hand with a time standard. Unlike other reset mechanisms, the second hand is connected neither to the winding stem nor the crown. Consequently, when the crown is pulled out, the balance remains in oscillation and the movement continues to run, significantly reducing material stress. Pressing a separate button on the side of the case at 8:00 activates the reset mechanism. Exquisitely finished components enhance the artistry of the Glashütte Original Caliber 100-02's design: the signature Glashütte three-quarter plate with Glashütte ribbing, skeletonized rotor with 21K gold oscillation weight, twin spring barrels and swan-neck fine adjustment.

GUY ELLIA

TIME SPACE QUANTIEME PERPETUEL

The Time Space, original ambassador of Guy Ellia, is transformed into a perpetual calendar for this version. A mechanical hand-winding Frédéric Piguet caliber harbored in a 46.8mm black-gold case allows the owner to read the hour, minute, day, month, date and moonphases, while taking leap years into account. With a brushed bottom plate of 7.75mm, this watch symbolizes elegance, complexity and thinness. A specific automatic winder is integrated to the watch box. This model is also available in white or pink gold.

GUY ELLIA

TIME SPACE QUANTIEME PERPETUEL – FULL SET

The case and dial of this extraordinary piece are covered in pavé-set diamonds. A mechanical hand-wound Frédéric Piguet caliber harbored in a 46.8mm white-gold case allows the owner to read the hour, minute, day, month, date and moonphase, while taking leap years into account. With a brushed bottom plate of 7.75mm, this clear allusion to the Couteaux watches symbolizes elegance, complexity and thinness. A specific automatic winder is integrated into the watch box and this model is also available in pink gold.

H. MOSER & CIE.

MOSER PERPETUAL GOLDEN EDITION – REF. 341.101-008

In honor of the fifth anniversary of H. Moser's Perpetual 1, the house introduced the Perpetual Golden Edition in a strictly limited edition of 100 pieces. As the name of this model suggests, gold can be found in several parts of this special watch. In most Moser watches, the pallet fork and escapement wheel are made from solid gold, but in this timepiece, gold comprises the double hairspring escapement, the plate and bridges, and of course the dazzling three-part rose-gold designer case. The perpetual calendar can be adjusted forward and backwards at any time, and changes in less than one second—even when it advances directly from February 28th to March 1st—representing watchmaking at its finest.

H. MOSER & CIE.

MOSER PERPETUAL 1 – PALLADIUM – REF. 341.101-009

Complete with a hand-wound movement and power reserve of 7 days, the Moser Perpetual 1 in Palladium has a double hairspring escapement with two Straumann hairsprings to prevent gravitational error in combination with the house's well-known perpetual flash calendar. The functions of the calendar include a large date window at 3:00, a month display from the center via a hand, and leap-year display on the back that is adjustable via a pin pusher at 9:00. All flash calendar functions change in under one second, including adjusting the calendar forward or backward via the crown. This model comes with a hand-crafted fumé dial and a three-part palladium designer case.

IWC

SPITFIRE PERPETUAL CALENDAR DIGITAL DATE-MONTH – REF. IW379103

Powered by the automatic-winding 89800 caliber, this model comes complete with a flyback and stopwatch function, a totalizer at 12:00 and a unique perpetual calendar. The month at 3:00, date at 9:00 and leap-year indicator at 6:00 are all displayed digitally, a first for IWC. These functions are presented in large, legible Arabic numerals and pay homage to the clarity of the altimeter in a pilot's cockpit. Presented in a 18K red-gold case on a brown alligator leather strap, the striking and sophisticated Spitfire Perpetual Calendar Digital Date-Month has polished and satin finishes that complement the bold dial.

IWC

PORTUGUESE TOURBILLON MYSTERE RETROGRADE – REF. IW504401

With an appearance as magical as it is distinctive, the Portuguese Tourbillon Mystère Rétrograde is guaranteed to attract inquisitive glances. Watch lovers will be particularly fascinated by the unusual arrangement of the flying tourbillon, consisting of 81 parts, against a deep black background, creating the illusion that the filigree cage containing the balance is rotating in mid-air. Set in a mirror-finished ring, it forms the optical centerpiece of the entire dial. The retrograde date display is not only an original complication but also makes a good deal of sense, because a conventional date disc would conceal the tourbillon. After the 31st of the month, it automatically jumps back to the 1st; in shorter months, the hand can be rapidly advanced until it reverts to the 1st. On the right-hand side of the dial, the seven-day power reserve display indicates how much energy remains in the IWC-manufactured 51900 caliber. As befitting a timepiece of this quality, the Portuguese Tourbillon Mystère Rétrograde premieres in a glamorous platinum case with a gray dial and in 18K rose-gold with a dial in silver.

IWC

PORTUGUESE PERPETUAL CALENDAR – REF. IW503202

The moon is useful to sailors on the open sea not only for navigational purposes. Its influence on coastal tides has always been of greater importance, because the timing of their ebb and flow is reliably dictated by the moon: at new and full moons, high tides are exceptionally high and low tides exceptionally low. In the English Channel the difference can be up to 11.5m and in the Gulf of Maine as much as 21m, which illustrates the enormous importance of the moon for shipping. Aside from the date, day, month and year in four digits, the Portuguese Perpetual Calendar also indicates the number of days remaining until the next full moon. The display showing its course and featuring mirror images of the moon in the Northern and Southern Hemispheres deviates from the moon's actual progress by just 12 seconds in one lunar period.

IWC

PORTUGUESE PERPETUAL CALENDAR – REF. IW502306

The moonphase display on the Portuguese Perpetual Calendar, Reference 5023, presents grand-scale theatrics on a tiny stage. This elegant, up-to-the-minute timepiece leaves virtually no wish unfulfilled: perpetual calendar, four-digit year display, seven-day automatic movement with Pellaton winding system and a power reserve display. With a power reserve of seven days, the in-house manufactured movement on this watch beats at 21,600 vph. This model comes with a shining 18K rose-gold case and a silver plated dial or in a 18K white-gold case with a dial in a deep slate hue.

PARMIGIANI

TONDA CENTUM – REF. PFH227-1200300-H1441

With its peripheral hand-driven retrograde date and unique revelation of the perpetual calendar's dynamic components, this 42mm 18K white-gold timepiece blends the optimal expression of sophisticated timekeeping with an ingenious demonstration of haute horology's mechanical harmony. Driven by the self-winding PF 333 caliber, and worn on an Hermès alligator leather strap, the individually numbered wristwatch indicates the hours, minutes and central seconds, along with a comprehensive perpetual calendar, whose precision moonphase is accurate to one day every 122 years.

PATEK PHILIPPE

CELESTIAL WITH DATE – REF. 6102P

This universe-inspired timekeeping expression is powered by the self-winding 240 LU CL C caliber with off-centered mini-rotor in 22K gold. Made up of three metallized sapphire crystal discs, the sky-chart dial depicts, using a framing ellipse, the portion of the sky visible from Geneva. With its brilliant and sophisticated use of space, the 44mm platinum timepiece displays the moonphase through an aperture that dynamically tracks the orbital position of the moon in the night sky. Indicating the hours and minutes of mean solar time, as well as the date via a third hand against the peripheral scale, the Celestial with Date looks to the stars with extraordinary precision for the spectacular telling of time.

PATEK PHILIPPE

MEN'S ANNUAL CALENDAR – REF. 5396/1R

This 38mm rose-gold timepiece with brown sunburst dial takes on the annual calendar concept with elegance and optimal architecture. While the day, date and month of the calendar appear via three legible apertures in the face, the moonphase is beautifully enclosed within a circular 24-hour subdial for optimal am/pm disambiguation. Driven by the 324 S QA LU 24H automatic caliber with Gyromax balance, three central hands depict the hours, minutes and sweeping seconds. The Men's Annual Calendar is worn on a gold bracelet with fold-over clasp.

ROGER DUBUIS

LA MONEGASQUE PERPETUAL CALENDAR – REF. RDDBMG0006

This miniature mechanization of time is presented in a 44mm pink-gold case with a titanium bezel coated in black DLC. With its 4Hz frequency of oscillation (28,800 vph), the 364-component RD821J caliber, with a 48-hour power reserve, is adjusted in five positions as it delivers the comprehensive indications of this perpetual calendar. Upon the rhodium-plated satin sunburst dial, this timepiece contains a lower half composed of a snailed date counter and seamlessly integrated moonphase. On the upper half, two gold windows set the frame for the French day of the week, as well as the month. Completing the nature of the calendar, a leap year indicator appears discreetly at 3:00. Finished with pink-gold applied Arabic numerals, this watch features a sapphire crystal caseback.

Moonphase

Girard-Perregaux
Vintage 1945
XXL Large Date
Moon-Phases

JAEGER-LECOULTRE
QUANTIEME LUNAIRE
JAQUET DROZ
SUN
NOV
To live in harmony with the moon,
A. LANGE & SÖHNE
GLASHÜTTE I/SA
BVLGARI
DANIEL ROTH
CAL 2300

you no longer need to buy a costly perpetual calendar or astronomical timepiece—or to discreetly consult the watch of your feminine companion. Today, the moonphase display has taken hold as a complication in its own right—on men's as well as women's watches. The moon has left its discreet aperture to play prima donna in fantastical scenes. We are also seeing the rise of the third dimension, with small lunar globes waxing and waning upon the dial.

The observation of the moon's cycle has always constituted—along with the course of the sun, the return of the seasons and the alternating of day and night—one of the primary foundations of calendar systems. This cycle famously inspired the ancient Egyptians to divide the year into twelve months, a system we inherited via the Greeks and Romans. Despite the near-universal adoption of the Gregorian calendar, essentially based on the course of the sun, the ancient lunar or luni-solar (i.e. based on both the sun and moon) calendars retain their full importance in certain cultures, and are notably used to determine the date of traditional or religious holidays, or for astrological calculations. This is the case, for example, in the Chinese and Jewish luni-solar calendars. Such is also the case for the Muslim calendar, in which the year contains twelve lunar months of 29.53059 solar days—adding up to about 11 days fewer than the Gregorian calendar.

In the West, the moonphase display was an essential component of the first great clocks of the Middle Ages. It plays a role on the clock of Saint-Etienne Cathedral in Bourges, in France (1424), and on Prague's Old Town Hall (15th century). It is featured on the dials of the first pocket watches, from the 16th century—well before watchmakers could present the minute or the second. The moonphase later became an inescapable element of watches devoted to astronomical, astrological or calendar indications—especially the famous perpetual calendars. In the 20th century, although the moon had lost much of its magnetic force over the human mind, watchmakers took it upon themselves to miniaturize the mechanism to the wristwatch format, while conserving a maximum of precision.

TECHNIQUE

HOW DOES IT WORK?

The moonphase is generally displayed via a mobile disc on which are depicted two perfectly round moons. A specially shaped window, with two rounded cutouts representing the "dark" side of the moon, allows the display to represent the diverse aspects of the heavenly body throughout its cycle (waxing, full, waning and new moons). The indicator changes once a day, as a rule.

A VARIABLE LEVEL OF PRECISION

Though they are all based on the same principle, not all window moonphase displays offer the same level of precision. Their exactitude depends to a large extent on the number of teeth on the moonphase disc. The length of one synodic, or lunar, month (the length of time from one new moon to another) is actually 29 days, 12 hours, 44 minutes and 2.8 seconds, or 29.53 days—making it impossible to simply multiply any units from solar civil time. The more teeth the disc has, the more precise the moonphase will be.

ORDINARY MOONPHASES

On ordinary moonphases, the disc bearing the two moons is linked to a wheel with 59 teeth, corresponding to two lunar months (2 x 29.5 days). This system falls behind 44 minutes and 2.8 seconds per month, or close to nine hours per year, a discrepancy that reaches a full day after two years, seven months and around 20 days.

ASTRONOMICAL MOONPHASES

High-end watches are equipped with a more complex device, called "astronomical moon," with a 135-toothed wheel. With this system, the lag between the mechanism and the actual lunar cycle comes to just one day after 122 years and 44 days. Certain recent timepieces have surpassed even this feat with impressive displays of precision (see the "Precision Records" section).

DOUBLE MOONPHASES

Some timepieces feature a simultaneous indication of the moonphase in both hemispheres. This particular feature is generally not a technical advance in and of itself (since it requires only revealing the two moons already present on most mobile discs), but it nonetheless gives the dial a charming touch of originality.

THE MOON'S AGE

Moonphases are sometimes completed by the indication of the moon's "age," i.e. the number of days that have passed since the last new moon.

Calendars with Moonphase

The moonphase indication has long been an almost inescapable element on complete or perpetual calendars (see chapter on Calendars). On these pieces, it most often occupies a small window at 6 o'clock. However, numerous brands have now chosen to reinterpret this tradition with more creative designs, moons at 12 o'clock, oversized windows or off-center displays.

Chopard, *L.U.C Lunar One*

The perpetual calendar L.U.C Lunar One from Chopard offers an "orbital" moonphase: the small round window, in which the silver moon waxes and wanes, itself rotates around the small seconds axis on a disc that simulates the star-spangled night sky.

Harry Winston, *Premier Perpetual Calendar*

On Harry Winston's Premier Perpetual Calendar, the moonphase shines at 6 o'clock on the off-center hours/minutes subdial, while the retrograde displays of the month and date compose a perpetually moving arrangement.

Dubey & Schaldenbrand
Grand Dôme DT
In designing the dial of the Grand Dôme DT, Dubey & Schaldenbrand arranged multiple displays—date, day, month, chronograph counters and moonphases—on a vertical axis, creating a highly original geometry.

Breguet, *Classique 7337*
Breguet's Classique 7337 associates an off-center hour/minute subdial with two windows for the date and day. The moonphase window at 12 o'clock also displays the "age" of the moon, i.e. the number of days since the last new moon.

Epos, *3391 Moonphase*
On a midnight blue dial sprinkled with stars, the 3391 Moonphase from Epos displays—in an unusual configuration—a moonphase, along with windows for the day and month, and a hand for the date.

Jaeger-LeCoultre, *Duomètre à Quantième Lunaire*
The Duomètre à Quantième Lunaire, from Jaeger-LeCoultre, with the Dual-Wing movement, features a moonphase window complemented by two hands, one for the moon's "age," the other for the moonphase in the Southern Hemisphere.

Pequignet, *Rue Royale*
Equipped with a manufacture-made caliber, the Rue Royale model from the French brand Pequignet displays the moonphase in a large vertical window—complemented by a double day/date window and a power reserve indicator.

Longines, *Master Collection Retrograde Moon Phases*
On the Master Collection Retrograde Moon Phases from Longines, the large moonphase window at 6 o'clock is complemented by four retrograde hands for the day, the date, small seconds and a 24-hour additional time zone.

The Moon is the Star

Over the last several years, the moonphase indication has been breaking free of the perpetual or complete calendar "straitjacket", and establishing itself as a complication in its own right. On certain timepieces, it is even the main attraction—with displays that can be quite poetic.

Zenith
El Primero Chronomaster Open Grande Date Moon & Sunphase

Using an ingenious system of superimposed discs, the automatic chronograph El Primero Chronomaster Open Grande Date Moon & Sunphase from Zenith displays not only the phases of the moon (at night), but also the course of the sun (during the day).

Graham, *Geo.Graham The Moon*

The Geo.Graham The Moon from Graham combines a flying tourbillon with a retrograde moonphase display that needs to be corrected by only one day in 122 years. The openworked dial and the movement are set with 45 diamonds representing the constellations.

Perrelet

Grande Phase de Lune centrale

Perrelet gives pride of place to the moonphase, with an aperture that occupies the entire top half of the dial as well as an indication of the moon's age.

MB&F

Horological Machine N°2-SV – Final Editions

Equipped with a rectangular case and two juxtaposed dials, the very avant-garde Horological Machine N°2-SV – Final Editions, from the watchmaking collective MB&F, uses the leftmost dial to display a double moonphase that refers to both the Northern and the Southern Hemispheres.

Louis Moinet, *Variograph*

On its Variograph model, Louis Moinet has opted to display, not the course of the moonphase, but the exact day of the full moon in the window at 6 o'clock.

Maurice Lacroix

Pontos Décentrique Phases de Lune

In the large crescent-shaped cutout on the Pontos Décentrique Phases de Lune from Maurice Lacroix, two superimposed discs reveal the moonphase and day/night indication. The design is daring and boldly asymmetrical.

Ulysse Nardin, *Moonstruck*

A worthy heir to Ulysse Nardin's astronomical watches, the Moonstruck continuously indicates the position of the moon and sun relative to any point on the globe; it thus allows its wearer to read the current moonphase and the state of the tides wherever one may be.

Hermès

Arceau Pocket Moonphase Retrograde Calendar

Powered by a movement from Vaucher Manufacture Fleurier, this Arceau Pocket watch from Hermès displays a large moonphase between 5 and 6 o'clock, as well as a retrograde date with central pointer. It is available in rose or white gold.

Patek Philippe

Celestial Ref. 5102

Equipped with a mobile map of the starry night sky, the highly complex Celestial from Patek Philippe provides a perpetual view of the exact configuration of the night sky over the Northern Hemisphere, complete with the visible movement of the stars and the moonphases, as well as the moon's position in the sky.

Girard-Perregaux
Vintage 1945 XXL Large Date Moon-Phases

Available in steel or in rose gold, the Vintage 1945 XXL Large Date Moon-Phases from Girard-Perregaux houses an automatic-winding manufacture-made movement in an elegant, rectangular, Art Deco-inspired case.

Bulgari, *Grande Lune*

The Bulgari Grande Lune, in the Daniel Roth collection, stands out for the original configuration of its two-tone dial, with an asymmetrically positioned oversize moonphase aperture and a pointer-type date display.

Chopard, *L.U.C Lunar Twin*

The L.U.C Lunar Twin from Chopard displays the moonphase in both hemispheres, with a degree of precision amounting to a one-day lag every 122 years, upon a large, round, off-center dial. Thanks to its two barrels, the automatic movement has a 65-hour power reserve.

Round Apertures

Alongside traditional apertures, we are also seeing an increasing number of round apertures, often equipped with mobile covers that represent the dark side of the moon. This enables horologers to create majestic extra-large and even extra-extra-large displays.

Jaquet Droz, *Eclipse Onyx*

On the Eclipse Onyx from Jaquet Droz, a small mobile black disc reproduces the moon's cycle by hiding and revealing a smiling moon, inspired by historical engravings. This spectacle is surrounded by eight applied stars.

MB&F, *MoonMachine*

MB&F gave free rein to the Finnish watchmaker Stepan Sarpaneva to create a special version of his HM3 Frog watch. This has resulted in the highly original MoonMachine, with a moonphase display via a round, crown-shaped aperture and a rotor that reproduces the constellations of the Northern Hemisphere.

Piaget, *Emperador Coussin Moonphase*

On Piaget's Emperador Coussin Moonphase, the disc representing the moon is a plaque of white gold that is heated to create craters such as those found on the face of the moon. Its hand-crafted production renders each watch one of a kind.

Andersen Genève, *Kamar*

For its Kamar model ("kamar" means "moon" in Arabic), Andersen Genève has created a round aperture equipped with a mother-of-pearl disc bearing both a dark and a light moon, all on a dial of aventurine or guilloché 22-karat blue gold.

Martin Braun, *Selene*

On its Selene watch ("selene" is the word for "moon" in ancient Greek), Martin Braun has opted to meld poetry and realism; a large moon based on a photograph is alternately hidden and revealed by a mobile cover.

Pierre Kunz, *A014 HMRL Tahiti Moon*

On the A014 HMRL Tahiti Moon from Pierre Kunz, with retrograde hour and minute displays, the moonphase is displayed in the center of a large oval decorated with metal stars and black mother-of-pearl from Tahiti.

Three-Dimensional Displays

A handful of watchmakers—most often independent artisans—have chosen to supplant the classic aperture systems with innovative three-dimensional displays. In these, the moon is depicted by a small mobile sphere (or a stationary sphere equipped with a mobile cover) reproducing the different faces of the moon throughout its cycle. This leads to very evocative stagings.

Christiaan van der Klaauw
Real Moon 1980

A three-dimensional moon takes pride of place on Christiaan van der Klaauw's Real Moon 1980. The astronomical indications are completed by a subdial that displays the height of the sun in relation to the horizon, as well as a hand at 3 o'clock signaling the eclipses of the sun and the moon.

Konstantin Chaykin, *Lunokhod*

On the Lunokhod from Russian watchmaker Konstantin Chaykin, the heavenly body is represented by a stationary sphere sculpted in wootz steel, an Indian metal whose granular quality recalls the lunar surface. The moonphase is displayed by means of a semi-circular cover in black rhodiumed silver, which rotates around a small globe.

De Bethune, *DB28*

The DB28 from De Bethune presents an original, patented system with a mobile sphere. The two hemispheres, in steel and platinum, are flame blued; the steel oxidizes and changes color, while the platinum maintains its natural hue.

Cyrus, *Klepcys*

The Klepcys, designed by Cyrus in collaboration with the watchmaker Jean-François Mojon, boasts a black disc that slowly rises to hide a spherical moon, dotted with craters. During the new moon, the orb gives way to the brand's logo.

Thomas Prescher
Mysterious Automatic Double Axis Tourbillon

On the Mysterious Automatic Double Axis Tourbillon from Thomas Prescher (whose movement is entirely hidden within the bezel), the two cylinders for the hours and minutes frame a spherical moonphase display.

Records for Precision

On ordinary moonphase watches, the complication will eventually show a one-day discrepancy after two years, seven months and around 20 days. On the moonphase complications within the luxury watches at the high end of the market, that difference has been reduced to one day over 122 year and 44 days. A few great brands have opted to face this technical challenge and develop moonphases that are much more precise—even resulting in no lag at all.

A. Lange & Söhne
1815 Moonphase
The German company A. Lange & Söhne has equipped its 1815 Moonphase with a gear train fitted with a particular gear-train ratio that guarantees a difference of just 6.61 seconds per lunar month, or just one day every 1,058 years!

Arnold & Son, *True Moon*
True Moon, from Arnold & Son, claims to provide a precise reproduction of the waxing and waning of the moon, adhering to the exact duration of synodic periods.

H. Moser & Cie.
Moser Perpetual Moon
The Perpetual Moon from H. Moser & Cie would need to run 1027.3 years to differ from the real lunar cycle by one full day—thanks to a continually revolving display linked to the hours mechanism (as opposed to most moonphases, which advance once per day). The vertical lines marking the eight quarters of the moon, fixed by international standards, allow an ultra-precise reading.

IWC, *Portuguese Perpetual Calendar*
Using an innovative system comprised of thee toothed wheels, the moonphase indication on IWC's Portuguese Perpetual Calendar has a discrepancy of just one day over 577 years. The double hemisphere display for the Northern and Southern Hemispheres is completed by two countdowns signaling the number of days until the next full moon.

The Feminine Side of the Moon

The moonphase display remains one of the favorite complications of ladies' watches, though it is still more widespread on masculine wrists. Catching the wave of this longstanding interest, both large brands and independent watchmakers offer a beautiful palette of quite refined ladies' models, most often adorned with diamonds and mother-of-pearl.

Corum
Admiral's Cup Legend 38
Mystery Moon

A magical ballet: as it spins steadily around, the finely worked mother-of-pearl dial upon the Admiral's Cup Legend 38 Mystery Moon from Corum moves the date aperture (which features the sun) as well as the moonphase across the dial.

Breguet

Heritage 8860 Phases de Lune Retrograde

Equipped with an elegantly curved tonneau-shaped case, a guilloché-centered dial and an automatic movement featuring a flat silicon balance-spring, the Heritage 8860 Phases de Lune Retrograde from Breguet displays the lunar cycle via a retrograde indicator at 12:30.

Glashütte Original, *PanoMaticLuna*

Resolutely asymmetrical, the PanoMaticLuna from Glashütte Original displays the moonphase in a large vertical aperture at 2 o'clock, which also indicates the moon's "age"—above a large panoramic date characteristic of the German brand.

Patek Philippe

Diamond Ribbon Ref. 4968

Illuminated by a diamond spiral that wraps around the case, Patek Philippe's Diamond Ribbon Ref. 4968 features a highly precise moonphase (accumulating a one-day discrepancy every 122 years) upon its mother-of-pearl dial.

Saskia maaike Bouvier, *The 8 Moons*

Created by a woman for women, The 8 Moons timepiece from Saskia maaike Bouvier sports a starry night sky as a backdrop for eight spheres that show not only the current phase of the moon (at 12 o'clock), but also those for the seven following nights.

Zenith, *Star Moonphase*

The Star Moonphase from Zenith—equipped with the famous, ultra-thin, automatic Elite movement—features an imposing moonphase at 6 o'clock upon a dial punctuated by large blue Roman numerals. The cushion-shaped case is available in steel or rose gold.

BREGUET

CLASSIQUE MOON PHASES – REF. 7787BR.12.9V6

Radiating Breguet's signature style of classical elegance, this 39mm 18K red-gold timepiece boasts an exquisite hand-engraved dial with two perfectly positioned indications. Joining the off-centered power reserve scale on the lower half of the watch, a superb moonphase at 12:00 enlightens the wearer on both the age and phase of the lunar cycle. Meticulously finished with a hand-engraved 18K gold oscillating weight, the 591 DRL caliber is enhanced with a Breguet balance wheel and silicon straight-line lever escapement.

BREGUET

CLASSIQUE HORA MUNDI – REF. 5717BR.US.9ZU

A world-first in mechanical watchmaking, the Classique Hora Mundi's self-winding 77FO caliber takes the watch's wearer on an unparalleled voyage in timekeeping. Equipped with a groundbreaking synchronized module with reprogrammable mechanical memory wheel, the timepiece travels from one pre-selected city to the next with a single push of the crown at 8:00. Exquisitely embellished with a variety of refined techniques, the entirety of the globe-motif dial's indications—moonphase, day/night indicator, date and time via two open-tipped blued steel Breguet hands—instantly jumps to the specifications of the chosen destination.

DE BETHUNE

DB25 L STARRY SKY – REF. DB25LWS3V2

Upon a sky in mirror-polished grade-5 titanium with white-gold stars sparkling into infinity, this 44mm white-gold timepiece's spherical moonphase indication rises through the plane to offer the three-dimensional impact of its palladium and flame-blued steel construction. Fascinating in its celestial profundity, the hand-wound wristwatch is powered by the hand-crafted DB2105 V1 caliber with self-regulating twin barrel, silicon/platinum balance and silicon escape wheel, oscillating at a frequency of 28,800 vph with a power reserve of six days and adorned with beautiful Côtes de Bethune.

DE BETHUNE

DB25 S JEWELRY

A resolutely feminine interpretation of De Bethune's star-studded-sky masterpiece, this 40mm model endows its spatial complication, accurate to one day every 122 years, with a galaxy of precious stones. Beneath the curved blued steel hands, stars in white gold and 21 diamonds illuminate a dial pierced by the presence, at 12:00, of a spherical moonphase set with 44 diamonds and 44 blue sapphires. Driven by the hand-wound DB 2105 hand-wound caliber, with hand-decorated mainplate, and titanium bridge, the timepiece, housed within a drum-shaped white-gold case set with 61 baguette-cut blue sapphires, holds as much spectacular mysticism within as it does in its celestial exterior.

DE BETHUNE

DB28 SPECIAL EDITION

Discreetly positioned at 6:00, a platinum and blackened steel spherical moonphase indicator adds dimension to this special celebration of a timepiece awarded the illustrious Aiguille d'Or in 2011. Enriched by a skeletonized hour circle and zirconium floating lugs that permit an exceptional fit upon the wrist, the titanium timepiece boasts a subtle performance indicator at 3:00, between a blue and red dot. Joining the continuing revelation of the intricate hand-wound DB 2115 caliber, with self-regulating twin barrel, through the sapphire crystal caseback, the power reserve indicator gives the wearer an additional reason to admire the full brilliance of this 42.6mm timepiece.

DE BETHUNE

DB28 WITH GPHG LOGO – REF. DB28TIS8

Opening a window onto a movement awarded the 2011 Aiguille d'Or by the Jury of the Grand Prix d'Horlogerie de Genève, the grade-5 titanium timepiece's dial, skeletonized and revealing the superb craftsmanship of its delta-shaped mainplate and De Bethune "stripes," is completed at 6:00 by a spherical moonphase indicator in platinum and blued steel, which revolves on its axis with superb accuracy, losing just one day every 122 years. On the caseback, the DB 2115 caliber's six-day power reserve indicator makes a delightful appearance through a window in the pocket watch-inspired design.

de GRISOGONO

OTTURATORE

The understated clarity of the Otturatore's Clou de Paris cobbled dial is at the heart of this watch's conceptual genius. With the press of the stylish gold pushbuttons that surround the crown on the side of the case, the patented high-performance sequential display selector rotates its aperture 90° to reveal one of the four secondary functions that loom beneath the watch face. Like a racecar instantly accelerating to its top speed and immediately braking to a full stop, this innovative sequencer mechanism completes its module transition 15 times faster than the blink of an eye. To the casual observer, this watch possesses a charming, unassuming elegance. To the connoisseur, the Otturatore is a truly spectacular feat of unprecedented innovation and de Grisogono expertise.

F.P. JOURNE

OCTA LUNE

The Octa collection caters to a clientele looking for authentic, rare and innovative horology in an "easy-to-wear" mechanical wristwatch. The automatic-winding Octa Lune offers over 120 hours of chronometric power reserve, reminding us of ancient times when the full moon was the only source of light during the night. While today we no longer depend on moonlight for night vision, the moonphase indication remains one of the most poetic of all horological complications. The hours, placed on the right side of the dial, are easily and quickly read. The Octa Lune won "Men's Watch of the Year" at the 2003 Grand Prix d'Horlogerie in Geneva.

FRANCK MULLER

MASTER BANKER LUNAR – REF. 8880 MB L WGE

Franck Muller introduced the Master Banker line for businessmen who are constantly on the go. This model offers three different time zones that are adjusted via the crown so the wearer easily knows the time at the stock markets of his choosing around the world. The day/night display of the two time zones is located in the upper part of the dial next to the counters with a small conjuring trick to ensure clear legibility beyond comparison. The Lunar model is also complete with a large and artful moonphase indicator at 6:00.

FREDERIQUE CONSTANT

FREDERIQUE CONSTANT MANUFACTURE TOURBILLON MOONPHASE DATE 24H REF. FC-985ABS4H9

Powered by the sophisticated FC-985 self-winding caliber, completely developed in-house, this multi-complication timepiece presents a black Clous de Paris guilloché dial, through which the ingeniously balanced, individually numbered tourbillon rises to glory. Matching the 44mm 18K rose-gold case, the central hands are joined in their time-telling journey by two multi-function subdials that magnify the extreme precision of the watch. At 9:00, the 24-hour indicator features an internal day/night indicator for optimal coherence. Positioned at 12:00, the moonphase indicator is surrounded by a contrasting small date ring. Designed with a silicon escapement wheel and lever, the movement of this limited edition horological work of art is finely decorated with Côtes de Genève and perlage, as well as chamfered and rhodium-plated bridges.

GLASHÜTTE ORIGINAL

PANOMATICLUNAR – REF. 90-02-42-32-05

The Pano collection's signature asymmetric effect is easily appreciated with a glance at the new PanoMaticLunar, which features the characteristic off-center visuals. The hour/minute and small seconds subdials are aligned along a vertical axis to the left, while the Panorama Date and superb moonphase display are positioned to the lower and upper right, respectively. The stainless steel case frames a warm silver dial with silver hour indexes beneath blued steel hands, complemented by the moonphase display and characteristic Panorama Date. Within beats the manufacture's superb Calibre 90-02, a self-winding 47-jewel movement with a power reserve of 42 hours. The sapphire crystal caseback provides a clear view of the finely finished movement, including the typical Glashütte three-quarter plate with Glashütte ribbing, off-center skeletonized 21K gold rotor, blued screws and hand-engraved balance cock.

HARRY WINSTON

HARRY WINSTON MIDNIGHT MOON PHASE – REF. 450/UQMP39WL.W/D3.1

This oneiric inspired timepiece takes a romantic stance on the moonphase complication. Part of the Harry Winston Midnight Collection, this impeccably elegant watch features a crescent-moon shaped aperture that extends from 9:00 to 12:00. Within the window lies a moon and stars that appear from behind the branches of trees, paying homage to nature. This romantic décor is meticulously engraved and lavishly finished in contrasting tinted lacquers on the dial. Three brilliant-cut diamond hour markers match the scintillating diamond-encrusted case. This model reveals Harry Winston's dedication to elegance, beauty and meticulous craftsmanship.

HARRY WINSTON

HARRY WINSTON OCEAN LADY MOON PHASE – REF. 400/UQMP36RC.MD02.D3.1

A new piece from a classic house, the Ocean Lady Moon Phase fuses traditional timekeeping methodology with a sporty, bold look. This luxe, ultra-feminine watch dazzles with 57 brilliant-cut diamonds on the bezel. To emphasize the strong look of the watch, the case is slightly larger than that of the traditional female model at Ø 36mm. The moonphase motif sparkles with 195 diamonds on an elegantly crafted white mother-of-pearl dial. A rose-gold buckle, set with diamonds, adds a touch of opulence to the white rubber strap, completing the vibrant and luxurious look of the model.

LONGINES

THE LONGINES MASTER COLLECTION RETROGRADE MOON PHASES – REF. L2.739.4.71.3

The Longines Master Collection Retrograde Moon Phases is fitted with an exclusive self-winding caliber (L707.2) developed especially for Longines. With a diameter of 44mm, it shows the time, day, night and moonphase, as well as four retrograde functions. The silvered dial features 11 black Roman numerals and six blued steel hands. The stainless steel case has a transparent caseback and is mounted on a brown alligator strap with a folding safety clasp.

PATEK PHILIPPE

LADIES GRAND COMPLICATIONS – REF. 7140R

Measuring 34.1mm, this timepiece's rose-gold case is lavishly embellished by a bezel set with 68 diamonds and a prong buckle set with 27 diamonds. Powered by the ultra-thin 240 Q self-winding movement with 22K gold off-centered mini-rotor, the Ladies Grand Complication depicts, on its white opaline dial, the brilliance of its perpetual calendar in a space-maximizing three-subdial design. Much like the display at 3:00, where the month and leap-year indications are seamlessly combined, a counter at 9:00 illustrates the day and 24-hour indication with a similar two-hand construction. At 6:00, the moonphase, which blends perfectly with the overall color scheme of the watch, is encircled by the hand-indicated date.

PATEK PHILIPPE

MEN'S GRAND COMPLICATIONS – REF. 5940

On the grained cream-colored dial of this yellow-gold timepiece, the ultra-thin 240 Q caliber's perpetual calendar comes to life with an architecture of harmonious integration. At 3:00 and 9:00 respectively, the month and leap-year indications, and day and 24-hour indications, are effortlessly combined to accommodate the watch's abundance of information on a finely proportioned canvas. At 6:00, the moonphase and date extend the dual-function-counter structural theme, and further express the uncluttered identity of this five-hand perpetual calendar. The timepiece is worn on a hand-stitched matte chocolate-brown alligator strap and boasts an interchangeable full back and sapphire crystal caseback.

ZENITH

EL PRIMERO CHRONOMASTER GRANDE DATE – REF. 18.2160.4047/01.C713

An alliance of useful complications, this chronograph model measures 1/10ths of a second and shows the moon- and sunphases. The indication reflecting the double path of the sun and the moon is displayed by means of two superimposed discs. The transparent upper disc with two moons performs one rotation every 59 days, while the lower day/night disc makes one turn every 24 hours. Through a dial opening at 11:00, El Primero Calibre 4047 reveals the visually stunning vision of its beating heart.

ZENITH

HERITAGE ZENITH STAR MOONPHASE – REF. 22.1925.692/01.C725

This original cushion-shaped case houses the ultra-thin mechanical automatic Zenith Elite 692 movement, beating at a frequency of 28,800 vph and endowed with a 50-hour power reserve. The remarkably designed and supremely elegant dial gives functions a leading role, with small seconds appearing at 9:00 and a moonphase display at 6:00 entirely dedicated to the path of the heavenly luminary.

Diving

Richard Mille
RM 028

First invented in the 1950s, in step with the new popularity of water sports,

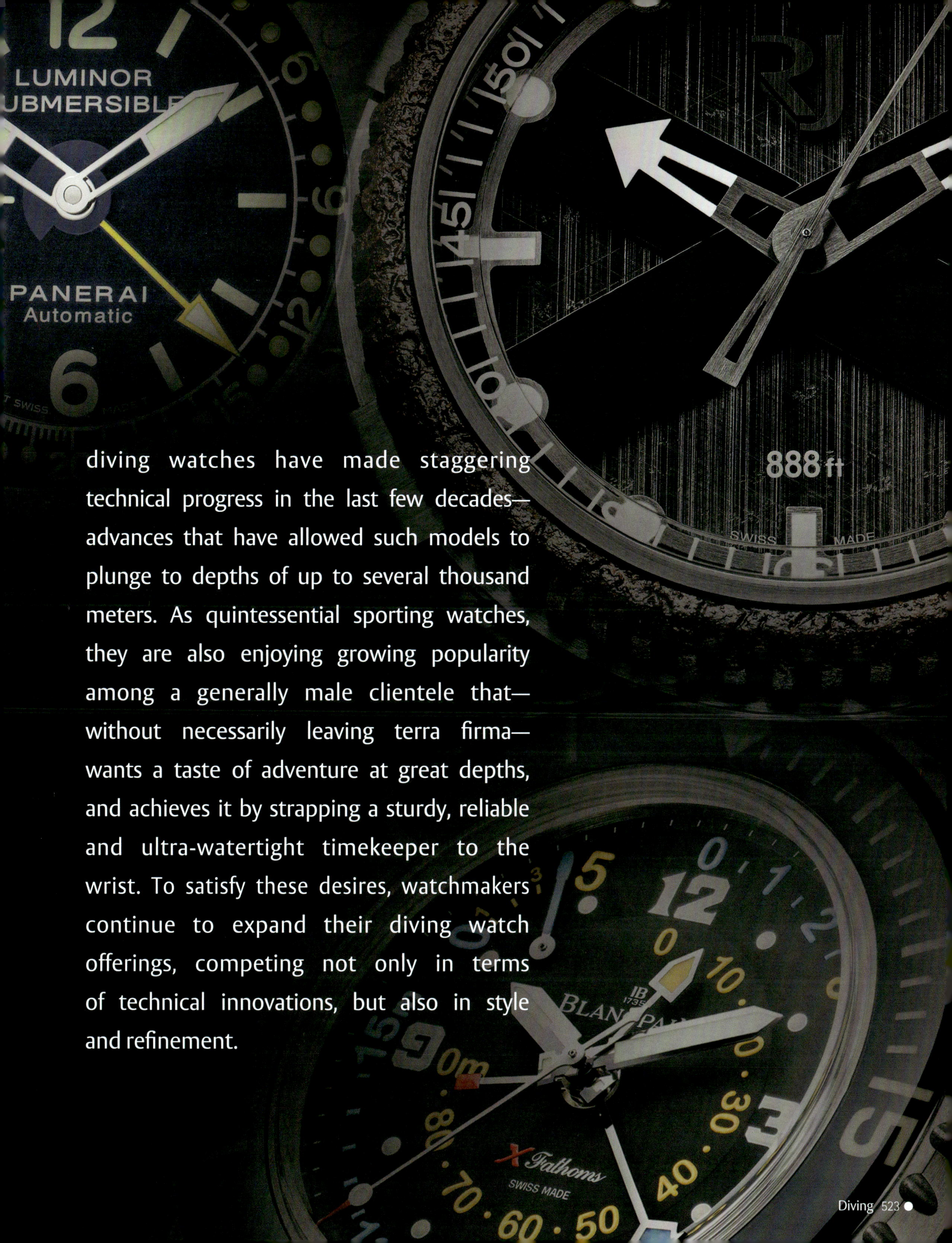

diving watches have made staggering technical progress in the last few decades—advances that have allowed such models to plunge to depths of up to several thousand meters. As quintessential sporting watches, they are also enjoying growing popularity among a generally male clientele that—without necessarily leaving terra firma—wants a taste of adventure at great depths, and achieves it by strapping a sturdy, reliable and ultra-watertight timekeeper to the wrist. To satisfy these desires, watchmakers continue to expand their diving watch offerings, competing not only in terms of technical innovations, but also in style and refinement.

Water resistance, as well as lubrication, has always constituted one of horologers' main concerns. This is not to say that all horological aficionados are budding Jacques Cousteaus. However, water resistance represents a welcome protection, even on terra firma, against the two great enemies of watch movements: humidity and dust.

The first water resistant watch, the famous Oyster from Rolex, arrived on the scene in 1926, thanks to the invention of the "Screw Crown." In 1953, the brand created for Professor Auguste Piccard a watch that would accompany his bathyscaphe to 3,150m under the sea. In 1960, the Deep Sea Special would descend to over 10,916m down in the Marianna Trench in the Pacific Ocean, and ascended to the surface intact after having undergone pressure of more than a ton per square centimeter.

Besides explorers and scientists, the first diving watches often equipped the armed forces. This was notably the case with Panerai in the 1930s and 1940s, made for the Italian Navy, with the models' luminescent indications on the dial and a crown system guaranteeing water resistance to 200m.

In the 1950s, with the rise of water sports, came the first collections devoted to diving watches. Rolex set the tone in 1953 with its famous Submariner, water resistant to 100m (now 300m). That same year, Blancpain launched its Fifty Fathoms, created in collaboration with two "frogmen" from the French Army. This watch, initially resistant to 50 fathoms of depths—hence its name—stands out for its black dial, bezel and strap. It would later be released in countless other versions.

Ever since, diving watches have become the quintessential sporting watch, and horologers have continued to increase their resistance and robustness.

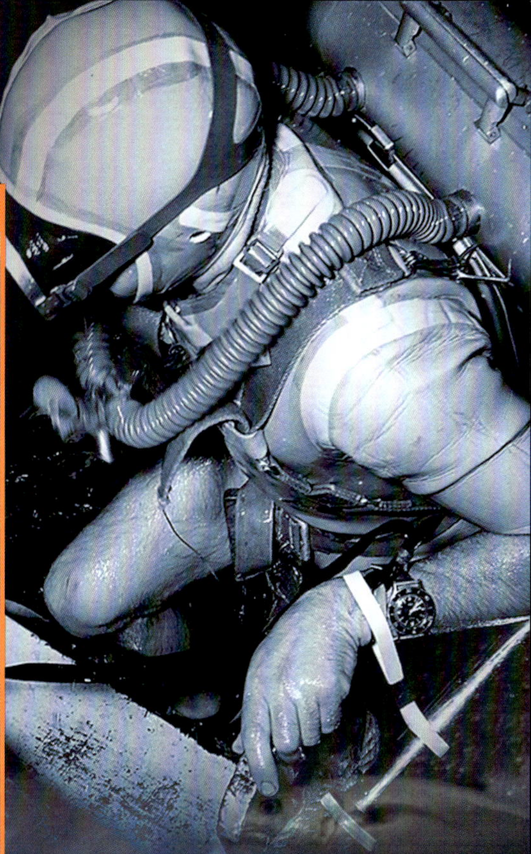

TECHNIQUE

HOW DOES IT WORK?

The definition of an authentic "diving watch" conforms to an official Swiss industry standard, or "norme de l'industrie horlogère suisse" (NIHS 92-11). To earn the title of diving watch, a timepiece must not just possess a water resistance superior to the average watch (minimum 100m), usually ensured by a screw-down crown and caseback. It must also offer all the functions that are indispensable to amateur or professional divers, notably a system that allows one to preset and check the time of the dive, with a device preventing any accidental manipulation. This is usually done by a turning bezel whose unidirectional rotation prevents the diver from inadvertently extending the remaining dive time.
The dial usually features large numerals and highly visible hands, generously covered in luminescent coating. The metal strap or rubber bracelet is often equipped with an extension system that stretches over the diving suit. The highest-performance models possess a valve that eliminates any helium within the watch. This highly volatile gas, used to facilitate breathing in pressurized chambers, gets inside the watch case very easily, but the case must be totally evacuated of helium when re-ascending to avoid the build-up of excessive pressure.

A SMALL REMINDER ABOUT WATER RESISTANCE

Contrary to what is all too often believed, the indication of water resistance in meters is a technical standard that does not correspond to an absolute value. The phrase "water resistant to 30m" does not imply that the watch can be worn to that depth under any conditions whatsoever. The movements of the wearer (dives, jumps, swimming, etc.) can increase the pressure considerably. For all aquatic activities, it is therefore best to stick to watches that possess a water resistance of at least 100m.

DIVING CHRONOGRAPHS

The inherently sporty nature of the two functions means that horologers often add chronographs to diving watches. However, it bears noting that most of these instruments, though perfectly "water resistant," cannot be activated in chronograph mode during the dive, or else water will seep into the case via the pushbuttons.

Re-Releasing Flagship Models

Several well-known brands are re-releasing or re-interpreting their emblematic diving watches, especially the pioneering models of the 1950s. Though their look might be similar to the original, their performances usually benefit from the enormous technical progress made in the intervening years.

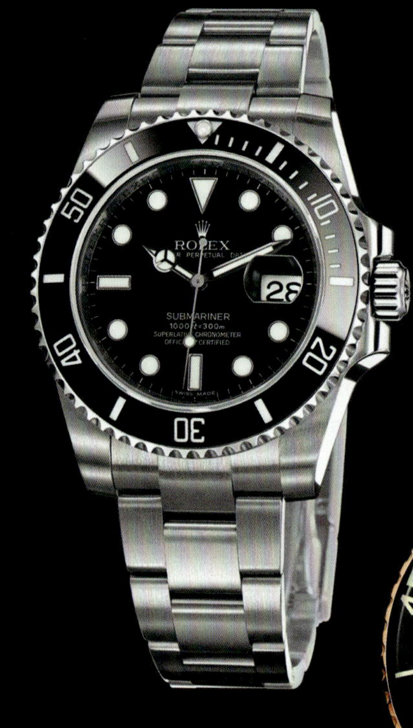

Rolex, *Oyster Perpetual Submariner*

Heir to the legendary 1953 model, the Rolex Oyster Perpetual Submariner is water resistant to 300m thanks to a patented Triplock screw-down crown system. The unidirectional rotating bezel is in ultra-resistant scratchproof ceramics.

Blancpain

Fifty Fathoms Flyback Chronograph 5085F-3630-52

Blancpain revisited its historic 1953 model Fifty Fathoms by creating a whole collection, notably including a flyback chronograph and water resistant pushbuttons that are functional to 300m.

Omega, *Seamaster PloProf 1200M*

Omega re-interpreted its 1970 Seamaster 600M model, also called PloProf (for "plongeurs professionnels," or professional divers), in a modernized version that is water resistant to 1200m, with a co-axial escapement and a security pushbutton that locks and unlocks the rotating bezel.

Tudor, *Heritage Black Bay*

Inspired by a 1954 model, Tudor's Heritage Black Bay model (water resistant to 200m) stands out for its "vintage chic" design, with a convex dial, red rotating bezel, "snowflake" hands and large luminescent hour markers.

Super Luminous

Diving watches must offer optimal readability in the extreme darkness prevailing at great depths. For this purpose, they are generally equipped with oversized numerals, indexes and hands that are treated with luminescent coating. Some brands have developed "lighting" systems offering even higher performance.

Ball Watch, *Engineer Hydrocarbon NEDU*

On Ball Watch's Engineer Hydrocarbon NEDU (water resistant to 600m), the hour markers and hands are equipped with micro-tubes filled with a luminescent gas. This Swiss technology guarantees a luminosity 100 times more intense than other methods.

Luminox, *Deep Dive*

The Luminox Light Technology System, in operation on the Deep Dive by Luminox, guarantees visibility in any conditions, thanks to its micro-tubes reflecting a blue light—the last color perceivable by the human eye after a certain depth.

Champions of Style

From sporty chic models destined mainly for poolside cocktails to professional instruments equipped for confronting the watery depths, from ultra-technical looks to timepieces focused on class and elegance, diving watches exist in a vast array of choices. Steel occasionally gives way to unusual materials—ceramics, titanium, bronze—and the quest for high performance is increasingly paired with close attention to style. These are watches that shine underwater, but also on dry land.

RJ-Romain Jerome, *Octopus*

Beyond its "rough" design integrating genuine pieces of the Titanic, RJ-Romain Jerome's Octopus (water resistant to 888 feet or 270m) is distinguished by its sapphire rotating ring that allows the wearer to regulate the dive time using the crown at 3 o'clock.

Ulysse Nardin
Black Sea Chronograph

Water resistant to 200m, the Black Sea Chronograph from Ulysse Nardin stands out for its totally black look with red and blue highlights, as well as its steel case with rubber coating. The unidirectional rotating bezel sports a wave motif.

Tudor, *Pelagos*

Water resistant to 500m, the Tudor Pelagos stakes its claim as a highly technical instrument, with a bezel equipped with a black ceramic disc, a helium valve and a satin-finished titanium bezel whose patina gives it a natural effect.

Omega

Seamaster Diver 300M James Bond 50th Anniversary

To celebrate the 50th year of the cinematic James Bond saga, Omega has released a limited edition of the watch worn by Agent 007 in five of the films. The Seamaster Diver 300M James Bond 50th Anniversary, equipped with a new co-axial escapement movement, is available in two sizes.

Richard Mille

RM 028 Brown PVD

Richard Mille has interpreted its RM 028 diving watch in an original, daring version, with a case in brown PVD and an orange strap. The skeletonized movement is equipped with a variable-geometry rotor that serves to adapt the winding to the wearer's activity.

Louis Vuitton
Tambour Diving II Auto Blue

The Tambour Diving II Auto Blue from Louis Vuitton possesses a rotating bezel with rubber lettering that facilitates handling. Another telling detail: the seconds counter evokes the red flag used by divers to signal their presence underwater.

Panerai
Luminor Submersible 1950
3 Days Automatic Bronzo

Water resistant to 300m, the Luminor Submersible 1950 3 Days Automatic Bronzo from Panerai stands out for its large case in bronze—a material often used on the old sailboats that the Italian-Swiss brand so admires.

JeanRichard, *Diverscope LPR*

The Diverscope LPR from JeanRichard possesses a rotating ring activated using one of the piece's two crowns. Its silhouetted large numeral 12 hosts an ingenious linear display system to indicate the power reserve, much like a fuel gauge.

Super Divers

Certain companies have chosen to compete by developing exceptional instruments capable of descending to 1,000m, 3,000m, 6,000m or even 12,000m underwater! These super divers possess ultra-robust and innovative architectures, reinforced water resistance systems and even—for the highest performers—cases filled with liquid that is immune to pressure.

Rolex, *Sea-Dweller Deepsea*

Rolex has reaffirmed its sovereignty over the deep seas with its launch of the Sea-Dweller Deepsea, water resistant to 3,900m. To resist such colossal pressures, this extraordinary instrument benefits from a patented new case architecture called Ringlock.

Girard-Perregaux, *Seahawk II Pro*

On its Seahawk II Pro, water resistant to 3,000m, Girard-Perregaux has shifted the crown to 4 o'clock to prevent it from getting in the way of the movements of the wrist, and equipped it with a large crown-guard that is both ergonomic and highly original.

Bell & Ross
Hydromax

Equipped with a quartz movement, the Hydromax from Bell & Ross is water resistant to the amazing depth of 11,000m, thanks to its case, filled with an oily liquid that is immune to pressure.

CX Swiss Military Watch™
20 000 FEET

The watchmaker CX Swiss Military Watch™ struck a blow for the record books with its 20,000 FEET, a mechanical timepiece that can plunge to more than 6,000m of depth.

Rolex
Oyster Perpetual Deepsea Challenge

Water resistant to 12,000m, Rolex's experimental diving watch Oyster Perpetual Deepsea Challenge accompanied director and explorer James Cameron in 2012 to 10,898m underwater in the Mariana Trench, the deepest solo dive in history.

Pita Barcelona, *Oceana*

Using two patented mechanisms that enable time setting by rotating the base and "reinforced magnetic gears," the Oceana model from Barcelona Pita—water resistant to 5,000m or 2,000m—can lay claim to being "the only watch that you can set underwater."

Breitling, *Superocean Chronograph M2000*

With a patented system of magnetic pushbuttons that operate through the case itself, the Superocean Chronograph M2000 from Breitling is presented as the only chronograph in the world that is watertight and functional at 2,000m underwater.

Doxa, *SUB 4000T Professional*

Water resistant to 4,000 feet (1,220m), Doxa's SUB 4000T Professional stands out by its orange dial, helium valve, "Safe Dive" power reserve indicator and its asymmetrical case, shaped to protect the crown.

Porsche Design, *P'6780 Diver*

On the P'6780 Diver from Porsche Design, water resistant to 1,000m, the automatic movement is lodged inside a pivoting chamber. When the chamber is locked, the rotating bezel and crown are immobilized. The user must unlock it to set the dive time.

Watches with Depth Gauges

Over the last few years, we have witnessed the return of a function that is not strictly horological, but very useful for divers: the depth gauge, or bathometer. It is a tempting alternative—or complement—to the electronic wristworn computers that are currently flooding the market.

IWC, *Aquatimer Deep Two*

The Aquatimer Deep Two from IWC is equipped with a mechanical depth gauge with two differently colored hands, one indicating the current depth (up to 60m), and the other the maximum depth achieved during the dive—a system that guarantees maximum safety.

Panerai

Luminor 1950 Submersible Depth Gauge

On its Luminor 1950 Submersible Depth Gauge, Panerai pairs an automatic caliber with an electronic depth gauge using a membrane and microchip system that activates a large yellow central hand. The levered bridge protecting the crown ensures water resistance to 120m.

Jaeger-LeCoultre

Master Compressor Diving Pro Geographic Navy SEALs

The Master Compressor Diving Pro Geographic Navy SEALs from Jaeger-LeCoultre stands out for its mechanical depth gauge equipped with a membrane that dilates or contracts according to the water pressure and then acts upon a large arrow-shaped hand. The depths measured range from 0 to 80m.

Blancpain, *X Fathoms*

Blancpain's X Fathoms houses a mechanical depth gauge that operates to 90m, with memory of the maximum depth achieved and a separate, ultra-precise (+/-30cm) display for the 0-15m zone, as well as a five-minute retrograde counter for the decompression stages.

AUDEMARS PIGUET

ROYAL OAK OFFSHORE DIVER – REF. 15706AU.OO.A002CA.01

The Royal Oak Offshore Diver is powered by Audemars Piguet's in-house, automatic-winding Manufacture caliber 3120 and is 42mm in diameter. The movement, comprised of 280 parts and beating at 21,600 vph, is enclosed in a forged carbon case with a black ceramic bezel, black rubber-molded crown and pushbutton. The dial features a rotating inner bezel ring bearing a graduated diving scale and luminescent coating on hands and hour markers. Its functions include dive-time measurement, hours, minutes, seconds and date. The case, water resistant to 300m, is mounted on a rubber strap with oversized titanium buckle.

BLANCPAIN

X FATHOMS – REF. 5018-1230-64

The X Fathoms revisits the characteristics of its iconic 1953 ancestor, combining them with a mechanical depth gauge to create the highest-performance mechanical diving watch ever produced. Depth measure up to 90m and maximum depth reached memory, separate indication on the 0-15m scale with exceptional +/- 30cm precision, retrograde five-minute counter for decompression stops—the X Fathoms concept watch abounds in world firsts. Its movement, reference 9918B, is based on the manufacture-made Calibre 1315, which powers several models in the Fifty Fathoms collection. Its imposing 55.65mm case made of satin-brushed titanium is water resistant to 300m. It features a helium decompression valve for saturation diving and the unidirectional rotating bezel characteristic to the collection for almost 60 years. The individual calibration of each X Fathoms watch guarantees maximal precision of depth indications.

HUBLOT

OCEANOGRAPHIC 4000M – REF. 731.NX.1190.RX

The product of 18 months of determined research, development and testing, the Oceanographic 4000M is nothing short of a groundbreaking watchmaking achievement. Tested to the depth of 5000m, this self-winding diving masterpiece is certified to an astonishing 4000m below the surface. Accounting for the extreme pressure of such depths, the synthetic sapphire crystal requires an extraordinary thickness to ensure watertightness. From a legibility standpoint, the 48mm titanium timepiece's dial components are optimized in size and treated with SuperLumiNova for underwater luminescence at points of intense darkness. Finished with a helium valve for the release of gases during ascent, the Oceanographic 4000M is worn on a rubber and nylon blended strap, sized longer than ordinary to be worn over a diver's protective suit.

IWC

AQUATIMER CHRONOGRAPH EDITION GALAPAGOS ISLANDS – REF. IW3767

The Aquatimer Chronograph Edition Galapagos Islands not only represents the harmonious development of IWC's of professional diver's watches, but is also the star attraction to mark a new partnership by IWC Schaffhausen and the Charles Darwin Foundation in aid of the environment, particularly on the Galapagos Islands. This model is an attractive, useful and extremely sporting symbol of this collaboration: black like the lava of the volcanic islands born from the ocean, and white like the clouds floating by above them. An intricate relief engraving of a giant tortoise on the back of the case, surrounded by the inscription "Tribute to the Charles Darwin Foundation – Galapagos Islands", represents the Charles Darwin Foundation and the Galapagos Islands.

IWC

AQUATIMER CHRONOGRAPH – REF. IW3769

The red gold version of the Aquatimer Chronograph demonstrates once again that elegance is not at variance with a highly functional sports watch at IWC Schaffhausen. Allowing the wearer to change directly from a wetsuit to a dinner jacket, this versatile model's 44mm case made of red gold guarantees that even the most stylish diver can plunge into the great depths of the sea. The rotating diving bezel is unlike any before as it features "background lighting." Pressed into the multi-component bezel is a 4-mm wide sapphire crystal ring, to which a concentrated coating of SuperLumiNova applied to its underside. The system is thus particularly suited to night dives making this watch extremely functional.

PANERAI

LUMINOR SUBMERSIBLE 1950 AMAGNETIC 3 DAYS AUTOMATIC TITANIO REF. PAM00389

Endowed with an innovative case construction that provides a staggering degree of magnetic resistance—40,000 A/m—this 47mm brushed-titanium diver's watch accompanies its wearer deep beneath the surface with subsidiary indications of date at 3:00 and small seconds at 9:00. Armed with a unidirectional bezel, the timepiece achieves its extraordinary anti-magnetic distinction through the housing of the P.9000 self-winding caliber within an internal protective case crafted of soft iron of the highest purity. Immediately beneath the dial, the Faraday cage crucially diverts the flow of undesired magnetic fields, ensuring the optimal operation of the timepiece's mechanical core.

PANERAI

LUMINOR SUBMERSIBLE TITANIO – 44MM – REF. PAM00025

Powered by a self-winding, exclusive Panerai OP III caliber, the Luminor Submersible Titanio beats at 28,800 alternations per hour with a power reserve of 42 hours. This superb diving model is complete with a brushed titanium unidirectional rotating bezel who turns in a counterclockwise direction. The bezel has a scale that is graduated for measuring the duration of immersions and ratchet click at minute intervals. The bezel matches the rest of the case, which is complete in AISI 316L brushed titanium. Panerai guarantees legibility for the wearer as the black dial has applied luminous hour markers. The rubber strap and adjustable buckle can be changed with a steel screwdriver giving the owner of this timepiece the option for a strap that is better suited for diving.

RJ-ROMAIN JEROME

METAL OCTOPUS – REF. RJ.T.AU.DI.003.01

The Octopus is RJ-Romain Jerome's very first diver's model. It stands out for both its aesthetic expression and its technical execution. The Octopus features two crowns, one at 9:00 to set the watch, and a second one at 3:00 for the elapsed time controller. It is equipped with an exclusive internal unidirectional rotating sapphire bezel system featuring engraved minutes graduation. The notched bezel of the Ø 45mm case contains oxidized and stabilized steel from the Titanic—another maritime giant. The Octopus is available in three different versions, each one in a limited edition of 888 pieces and water resistant to 888 feet (270m).

Brand Directory

A. LANGE & SÖHNE
Altenberger Strasse 15
01768 Glashütte
Germany
Tel: 49 35053 440
USA: 800 408 8147

AUDEMARS PIGUET
1348 Le Brassus
Switzerland
Tel: 41 21 845 14 00
USA: 212 758 8400

BELL & ROSS
8 Rue Copernic
75016 Paris
France
Tel: 33 1 73 73 93 00
USA: 888 307 7887

BLANCPAIN
Chemin de l'Etang 6
1094 Paudex
Switzerland
Tel: 41 21 796 36 36
USA: 877 520 1735

BREGUET
1344 L'Abbaye
Switzerland
Tel: 41 21 841 90 90
USA: 866 458 7488

CHANEL
25 Place du Marché St. Honoré
75001 Paris
France
Tel: 33 1 40 98 50 00
USA: 212 688 5055

CHRISTOPHE CLARET
Route du Soleil d'Or 2
2400 Le Locle
Switzerland
Tel: 41 32 933 80 80
USA: 954 610 2234

DE BETHUNE
Granges-Jaccard 6
1454 La Chaux L'Auberson
Switzerland
Tel: 41 24 455 26 00
Tel: 41 22 310 22 71

de GRISOGONO
176 bis Route de St. Julien
1228 Plan-les-Ouates
Switzerland
Tel: 41 22 817 81 00
USA: 212 439 4220

DIOR HORLOGERIE
44 Rue François 1er
75008 Paris
France
Tel: 33 1 40 73 59 84
USA: 212 931 2700

F.P. JOURNE
17, rue de l'Arquebuse
1204 Geneva
Switzerland
Tel: 41 22 322 09 09
USA: 305 572 9802

FRANCK MULLER
22 Route de Malagny
1294 Genthod
Switzerland
Tel: 41 22 959 88 88
France: 33 1 53 67 44 39
USA: 212 255 8499

FRÉDÉRIQUE CONSTANT SA
Chemin du Champ des Filles 32
1228 Plan-les-Ouates
Switzerland
Tel: 41 22 860 04 40
USA: 877 619 2824

GIRARD-PERREGAUX
Place Girardet 1
2301 La Chaux-de-Fonds
Switzerland
Tel: 41 32 911 33 33
France: 33 1 72 25 65 43
USA: 201 804 1904

GLASHÜTTE ORIGINAL
Altenberger Strasse 1
01768 Glashütte/Sachsen
Germany
Tel: 49 3 50 53 460
Franc e: 33 1 53 81 22 68
USA: 201 271 1400

GUY ELLIA
21 Rue de la Paix
75002 Paris
France
Tel: 33 1 53 30 25 25
NY: 212 888 0505

H. MOSER & CIE.
Rundbuckstrasse 10
8212 Neuhausen am Rheinfall
Switzerland
Tel: 41 52 674 0050

HARRY WINSTON SA
8 Chemin du Tourbillon
1228 Plan-les-Ouates
Switzerland
Tel: 41 22 716 29 00
USA: 212 245 2000

HAUTLENCE
Rue Numa Droz 150
2300 La Chaux-de-Fonds
Switzerland
Tel: 41 32 924 00 60

HUBLOT
Chemin de la Vuarpillière 33
1260 Nyon 2
Switzerland
Tel: 41 22 990 90 00
USA: 800 536 0636

IWC
Baumgartenstrasse 15
8201 Schaffhausen
Switzerland
Tel: 41 52 635 65 65
USA: 800 492 6755

JACOB & CO.
Chemin de Plein-Vent 1
1228 Arare
Switzerland
Tel: 41 22 310 69 62
USA: 212 719 5887

JAEGER-LECOULTRE
Rue de la Golisse 8
1347 Le Sentier
Switzerland
Tel: 41 21 845 02 02
USA: 212 308 2525

LONGINES
Saint-Imier 2610
Switzerland
Tel: 41 32 942 54 25
USA: 201 271 1400

PANERAI
Viale Monza 259
20126 Milan
Italy
Tel: 39 02 36 31 38
USA: 212 888 8788

PARMIGIANI
Rue du Temple 11
2114 Fleurier
Switzerland
Tel: 41 32 862 66 30
USA: 305 260 7770

PATEK PHILIPPE
Chemin du Pont du Centenaire 141
1228 Plan-les-Ouates
Switzerland
Tel: 41 22 884 20 20
USA: 212 218 1240

PIAGET
37 Chemin du Champ-des-Filles
1228 Plan-les-Ouates
Switzerland
Tel: 41 22 884 48 44
USA: 800 628 4344

RICHARD MILLE
11 rue du Jura
2345 Les Breuleux Jura
Switzerland
Tel: 41 32 959 43 53
USA: 310 205 5555

RJ-ROMAIN JEROME
Rue du Marché 11
1204 Geneva
Switzerland
Tel: 41 22 319 29 39

ROGER DUBUIS
1217 Meyrin 2 Geneva
Switzerland
Tel: 41 22 783 28 28
USA: 888 738 2847

TAG HEUER
Louis-Joseph Chevrolet 6A
2300 La Chaux-de-Fonds
Switzerland
Tel: 41 32 919 80 00
USA: 973 467 1890

VACHERON CONSTANTIN
Chemin du Tourbillon 10
1228 Plan-les-Ouates
Switzerland
Tel: 41 22 930 20 05
USA: 212 713 0707

ZENITH
2400 Le Locle
Switzerland
Tel: 41 32 930 62 62
USA: 973 467 1890